Turbocharging Normally Aspirated E

Turbocharging NA Engines on a Budget

- Theory
- Method
- Examples

By Robert "Greg" Wagoner

© 2012 Robert G. Wagoner
All rights reserved.

Turbocharging Normally Aspirated Engines on a Budget

By Robert "Greg" Wagoner

Third edition: October 2012

Published by Lulu Enterprises, Inc.

Lulu ID # 13160495

ISBN 978-1-300-17994-8

© 2012 Robert G. Wagoner

All rights reserved. No part of this publication may be reproduced, or transmitted in any form, by any means, electronic or mechanical, including photocopying, recording, or by any information retrieval system, without prior written permission of the copyright holder.

NOTICE: While every attempt has been made to ensure that the information in this manual is correct, no liability can be accepted by the author or publisher for loss, damage, or injury caused by any errors in, or omissions from, the information given. All recommendations on parts and procedures are made without any guarantees on the part of the author or the publisher. Author and publisher disclaim all liability incurred in connection with the use of this information.

PURPOSE OF THIS BOOK

"There is no replacement for displacement". However, when budget limited, installing a turbocharger is more cost effective than an engine swap, and turbocharging adds less weight to the car than putting in a larger engine. Win - win.

I have read about a dozen books on turbocharging. These books do a great job of explaining all the technical details of turbochargers and turbocharging theory. With such wonderful reference books available, I see no need to make another book like that.

I have also read a lot of articles on highly modified turbocharged engines with an ultimate goal of extremely high horsepower for racing applications. While these are certainly interesting to read about, they are beyond my budget. Hence, this book will not address expensive race engines either.

On the other hand, I have not found a book that gives a simple method of turbocharging normally aspirated engines at low cost and with reliability as a goal. It seems odd to me there is no such book, since a lot of people want more power and the "whoosh" sound of a blow-off valve, without spending a fortune. I am one of those people. I turbocharge just about every car I own for Autocross racing. Hence the purpose of this book is to explain how I turbocharge normally aspirated engines at low cost, and to provide numerous examples for a variety of different cars.

ACKNOWLEDGMENTS

I want to give credit where credit is due. The following list of people have helped and supported me on these projects and with the writing of this book. I owe them a huge debt of gratitude.

Nick Wagoner, my son - sometimes my student, sometimes my teacher, always my favorite companion to work together on projects. He helped with most of the projects in this book, and he contributed a significant portion of ideas and text to the book itself.

Bud Wagoner, my father - the man who first sparked my interest in automobiles, and who first taught me how to do body work and rebuild engines. All he asked in return was for me to pass the knowledge down to my son, like he did for me.

Terry Jones, my wife - who always loves and supports me even when I spend many late nights and weekends working on the projects in this book and this book itself.

Seth O'Dell, my friend - for letting me use his turbocharged Miata as an example in this book, and for providing valuable inputs as he reviewed the book before it was published.

© 2012 Robert G. Wagoner
All rights reserved.

Turbocharging Normally Aspirated Engines on a Budget

CONTENTS

Introduction	6
Don't do this to your daily driver	7
Skill level / Prerequisites	7
A Brief History of Turbocharging	7
1) Overview	9
HP/Weight Ratio	10
Supercharger and Turbocharger Overview	11
Definitions and Terminology	11
2) Turbocharging: Making It Reliable	17
Designing for Reliability of the Turbocharger Itself	17
Designing for Reliability of the Entire System	17
Designing for Reliability of the Engine: Detonation and Spark Knock	18
Reducing Compression Ratio to Avoid Detonation	18
Colder Plugs to Avoid Detonation	18
Adding an Intercooler to Avoid Detonation	19
Sizing Fuel injectors for Reliability	21
BSFC, Brake Specific Fuel Consumption	22
Blow-Off Valve (BOV)	22
Wastegate	23
Engine Mounts	24
Clutch	24
Tuning: Reprogramming the ECM	24
Camshafts for Turbocharged Engines	25
Start with a Solid Engine and Drivetrain	25
3) Example: Turbocharging a 3.4L V6 in a Pontiac Fiero	27
Turbocharger Selection	28
Using the Excel Spreadsheet for Turbocharger Compressor Sizing	29
Turbocharger System Block Diagram	30
Low cost parts and sources, new and used	31
Initial Build with the "Small" Turbocharger	31
Mandrel Bent Exhaust Tubing: Improved Flow at Higher Cost	33
Connecting Two Exhaust Pipes Together	33
Oil System	34
Intake and PCV	34
Manifold Absolute Pressure (MAP) sensors	35
The First Problem: Buckled Exhaust	37
Spark Knock	37
Colder Plugs	37
Reducing Compression Ratio	38
Fuel Injectors	38
Fuel Pump	39
Adding an Intercooler	40
More Torque Requires Stronger Support	40
Adding a BOV	41
New Gauges for Tuning	41
Reprogramming the Fiero ECM	41
The Recommended Configuration	45
Parts List and Costs	45
Breakage Cost	45

© 2012 Robert G. Wagoner
All rights reserved.

4) Example: Turbocharging a 1.8L 1ZZ-FED Engine in a 2001 MR2 Spyder — 47
Part 1: Turbocharger System Design — 48

- Turbocharger System Block Diagram — 48
- Mass Air Flow (MAF) Sensor Location — 48
- Cooling and Lubrication — 49
- Options Considered — 49
- Installing AEM FIC: Test Data to Decide on the Best Option — 51
- Initial Measurements — 53
- Test data on engine before installation of turbocharger — 54
- MAF Sensor Modification — 55
- Initial Calculation of expected HP — 56
- Installation of Turbocharger — 57
- The Turbocharger Itself — 58
- Initial Test of the Turbocharged Engine with No Compensation — 59
- Part 2: Tuning — 60
- AEM FIC Parameter set "FIC rev 5" — 61
- AEM FIC Parameter set "FIC rev 18" — 62
- AEM FIC Parameter set "FIC rev 55" — 63
- Piggyback FIC Bypass Option - Impact on Availability — 67
- Finding Upper Limit of Boost Pressure — 67
- Parts List and Costs and Breakage Cost — 69

5) Example: Turbocharging a 2.0L ECC DOHC Engine in a 1999 Dodge Neon — 71

- Fuel System Modification with 10:1 FMU — 72
- WARNING: High Fuel Pressure — 72
- Excel Spreadsheet Calculations of Airflow and Horsepower — 73
- Turbocharger Selection — 74
- Hardware Installation and Exhaust / External Wastegate — 74
- Turbocharger System Block Diagram and Ignition Timing — 75
- Cooling and Lubrication — 76
- Crankcase Ventilation and Crankcase Breather Problem — 76
- FMU Reliability — 77
- Parts List and Costs and Breakage Cost — 77
- Final Recommendation: Don't Use an FMU — 77

6) Example: Turbocharging a 3.0L 2JZ-GE Engine in a 1995 Toyota Supra — 79

7) Example: Turbocharging a 1.8L Engine in a 1997 Mazda Miata — 85

8) Example: Turbocharging a 2.8L V-6 Engine in a 1987 Pontiac Fiero — 91

9) Conclusions — 99

A) An Excel Spreadsheet for Turbocharger Compressor Sizing — 100
B) Explanation of Formulas used in the spreadsheet — 104
C) VE Calculations for Various Engines — 107
D) Turbocharging Carbureted Engines — 110
E) Turbocharging a 5.7L LS1 in a C5 Corvette — 111
F) Rebuilding Turbochargers — 114
G) Camshafts for Turbocharged Engines — 115
H) Superchargers vs. Turbochargers — 117
I) References organized by the author's preference — 119

Index — 120

Turbocharging Normally Aspirated Engines on a Budget

INTRODUCTION

I have been making cars go fast since I was old enough to drive. In my teens, my first racecar was a 1974 Nova that was born with a 350 small block, which I upgraded to a 454 big block, radical camshaft, headers, Edelbrock intake, 850 CFM Holley carburetor, and nitrous oxide for drag racing. Back then I raced on a ¼ mile dragstrip. I learned about the potential benefits of turbochargers while I was working on this Nova, so in 1982 I purchased a used turbocharger from a diesel truck engine for my Nova. I spent quite a bit of time plumbing it into the top of my Holley, only to find out that Holley was not designed to work under pressure, and the plumbing was too complicated for me at the time to put the turbocharger between the carburetor and the intake manifold. While that project was unsuccessful, that was the initial experience where I started learning about turbochargers. As Thomas Edison said, "I am not discouraged, because every wrong attempt discarded is another step forward." The first car I successfully turbocharged was a Pontiac Fiero. This is a car that was not really meant to be turbocharged. By that I mean turbocharging some cars is easier than others, and the Pontiac Fiero is not one of the easy ones. I wrote a book called "High Performance Fieros", which was given the thumbs up in the October 2007 issue of "Hot Rod" magazine. There is a chapter in that book about turbocharging the 3.4L V6 engine, and I have copied parts of that chapter into this book as the first example. Since I wrote that book, I have subsequently turbocharged three more Pontiac Fieros with the 2.8L V6.

To give some background on my Fiero experience, as soon as I graduated from college in 1985, I purchased my first new car, a black Pontiac Fiero GT. Since then I have owned about 12 more Fieros, many of which I upgraded in some way and sold, and the others were used for parts. I enjoy restoring them and making them perform better than original. My son, Nick, has also taken an interest in Fieros. At age 13, he bought 2 Fieros: a 1986 Fiero SE that ran poorly, and a 1987 Fiero GT that didn't run at all. He combined the two, taking the best parts from the 1987 Fiero GT to make one nice 1986 Fiero SE, and then he sold the 1987 body to recover enough money to continue the project. He chose the license plate "KWIK CAR" before it was actually a quick car, and together we have tried to make it live up to its name ever since. Nick's Fiero was featured in two articles on the internet: July 2005 of "Suncoast Fieros" and at "Pennock's Fiero Forum" in January 2006. After a body style change it was in the March 2006 issue of "Hot Rod" magazine.

More recently I spent about a year modifying and tuning a 2002 Subaru Impreza WRX trying to maximize the boost. After that I spent about a year turbocharging a 1999 Dodge Neon R/T. Next I turbocharged a 1995 Toyota Supra, which was actually easier than most of the others. Then I worked with my son to build up and turbocharge a hybrid EJ22T Subaru engine with high boost. Next I turbocharged a Toyota 2001 MR2 Spyder. Besides these and other projects that I have taken to completion, I also have designed a number of other potential projects on paper only, such as a turbocharged LS1 engine. In this book I will provide details of these projects and others as examples.

I make some recommendations in this book that are contrary to internet lore. For example, I use relatively small turbochargers, because I prefer to get the highest efficiency in the RPM range I operate the engine the most, and because this minimizes turbo lag, and finally because this keeps the cost lower. You will find refreshing new ideas in this book with reasoning to justify them, new ideas that are sometimes completely opposite to those of people building high-dollar race engines.

I hope you enjoy this book and find the information to be useful in your own projects.

– *Robert "Greg" Wagoner*

© 2012 Robert G. Wagoner
All rights reserved.

Turbocharging Normally Aspirated Engines on a Budget

Don't do this to your daily driver car

I only make such extensive modifications to a car I plan to autocross race. On every car I have modified, the custom parts are nearly impossible to get right the first time. Custom work requires trial and error, and besides the time it takes, there is also a "breakage" cost associated with it. Custom engineering a turbocharger system for a car will take a long time to get running right, and it will not be right the first time, and when finished it will not be as reliable as a stock engine.

>Don't do this to your daily driver car.

Skill level / Prerequisites

Tools: To reduce the cost of the work, I don't buy pre-engineered and pre-fabricated kits, but instead, each system is unique: I build a custom exhaust, for example. For these projects to be possible at the lowest cost, it takes some basic tools and skills, to avoid paying others to do the work. To cut the aluminum intercooler piping and steel exhaust requires some type of cutting tool, either a metal saw or body grinder with a metal cutting blade. To weld the steel exhaust pipes requires a MIG welder, or a friend with a welder. I splice the aluminum intercooler piping together with silicone connectors and clamps, so no aluminum TIG welding is required.

Ability: Custom engineering a turbocharger system for a car requires research, planning, calculations, hardware installation, and it usually takes a lot of time tuning to get it to run right. In this book I try to describe how we did this work in as much detail as necessary without getting into too much detail, such as the exact routing of every exhaust pipe. These are not kits, and each car will be a bit different. Even when I have turbocharged a number of different Fieros all with a V-6 engine, each time I changed and improved the design a bit, based on the "lessons learned" from the previous installation. For a person who enjoys a challenge, this is it.

If it sounds overwhelming, don't give up yet. Read on. Make your decision at the end of the book.

A Brief History of Turbocharging

The reference books listed in the back of this book can provide many details of the history of turbocharging. Here is a short history lesson: highlights that interested the author. To me it was very interesting to learn how long turbocharging technology has been around.

- 1885- An engineer named Gottlieb Daimler patented the first supercharger, which he described as a gear-driven pump that forces air into an internal combustion engine.
- 1905- An engineer named Alfred Büchi patented the first turbocharger. In the following decade turbocharged diesel engines were introduced into diesel ship and locomotive applications.
- 1918- An engineer named Sanford Moss turbocharged a V12 Liberty engine for aircraft use, which was tested at an elevation 14,000 feet at the top of Pike's Peak, Colorado, where it put out over 350 HP, compared to the roughly 220 HP the same engine could produce at this altitude without a turbocharger. This demonstration was the gateway for turbochargers onto aircraft engines.
- 1943- Turbocharged aircraft (such as the P-38, P-47, B-17, B-24, B-29, P-38, and P-51) were applied in mass numbers during WWII, primarily because the turbocharged engines had high power to weight ratio with excellent high altitude performance.
- 1962- General Motors introduced the first production cars equipped with a turbocharger: the Chevrolet Corvair Monza and the Oldsmobile Jetfire.
- 1982- Since this is my history lesson, the next important milestone in the history of turbocharging was 1982 when I bought my first turbocharger for use on a normally aspirated engine. ☺

> **Warning**
> Modifications in this book may not be legal for street cars. Check local and state laws before making any changes to your car.
>
> Modifying your engine will void any factory warranties.
>
> The author does not recommend that anyone make any modifications to their car. The author is not responsible for any loss, damage, or injury caused by any modifications anyone makes to their car.

Expect Breakage
- I have had multiple major failures when modifying engines to increase power.
- I have learned to start a project like this with an expectation of major damage, and I feel lucky if that doesn't happen.
- I have done the most damage when I decided at the end to bump up the boost more and more until I "find the limit".

Reliability
- Minimize the number of modifications – every change reduces reliability.
- Start with a good- running, low-mileage engine, clutch, and drivetrain.
- Ensure every modification is done carefully and correctly.
- Plan on a conservative power increase – don't push it.

> *This book should be used in conjunction with a factory service manual, or a repair manual for your specific vehicle from GM, Chilton, or Haynes, which will include safety procedures, engine rebuilding information, torque specifications, cleaning, etc.*

Turbocharging Normally Aspirated Engines on a Budget

1 Overview

There are many ways of spending money on an engine to improve performance, and this book focuses on getting the most benefit for the least money. This book gives numerous examples of turbocharging normally aspirated automobile engines using a method {or maybe more like a philosophy} of changing as little as possible to achieve a quite noticeable increase in performance. In general the method is simple… all mass production normally aspirated engines are designed with some margin in the fuel system, drivetrain, etc. because the design engineers know the margin is required to make an extremely reliable mass production engine with adverse environmental conditions (hot, cold, rain, altitude, etc.) and also considering some drivers really push these engines. This margin allows a turbocharger to be added to ANY normally aspirated engine with SOME level of boost pressure. Of course the boost pressure must be limited to keep the additional mechanical stress provided by turbocharging within the engine's capability and to avoid detonation (spark knock). In fact most normally aspirated engines have enough margin to operate with 7 PSI of boost, which is enough to increase the engine torque and horsepower by roughly 1.5 times. This naturally results in a low cost turbocharger installation, because most of the engine does not need to be replaced. By low cost, I mean in the neighborhood of $500 to $1000. This book does not describe any $5000 to $10,000 race engines. The method of turbocharging described in this book works great for a person on a budget and who wants a MUCH faster car.

This book provides specific examples of using this method of "turbocharging on a budget" on the following normally aspirated engines, each in its own chapter. These are organized so that each chapter teaches new concepts, and they tend to flow better if read in sequence. Consequently the latter chapters tend to get shorter and shorter, with the expectation that the reader has already learned certain concepts in previous chapters.

- Ch3 - 1986 Pontiac Fiero 3.4L V-6 engine (from ~153 HP NA to ~231 HP with 11 PSI boost)
- Ch4 - 2001 Toyota MR2 Spyder with 1.8L I-4 engine (from ~138 HP NA to ~207 HP with 9 PSI boost)
- Ch5 - 1999 Dodge Neon R/T with 2.0L I-4 engine (from ~150 HP NA to ~192 HP 6 PSI boost)
- Ch6 - 1995 Toyota Supra with 3.0L I-6 engine (from ~220 HP NA to ~306 HP with 8 PSI boost)
- Ch7 - 1997 Mazda Miata with 1.8L I-4 engine (from ~133 HP NA to ~205 HP 10 PSI boost)
- Ch8 - 1987 Pontiac Fiero with 2.8L V-6 engine (from ~135 HP NA to ~167 HP 8 PSI boost)
- App D - 2003 Chevrolet Corvette 5.7L V-8 LS1 engine (from ~350 HP NA to ~467 HP with 7 PSI boost)

The table on the next page compares these engines, before and after turbocharging, to other cars and ranks them in order of performance (HP / Weight Ratio).

This book includes a brief history of turbocharging and just enough theory to explain the how & why of this method. However, the history and theory can be found in more detail in many of the references listed at the end of this book. Repeating a lot of history or theory herein seemed to be redundant in light of these excellent existing reference books. Instead the real value of this book is that it focuses more on actual examples of how to turbocharge cars using this method, providing enough options and variations to allow a reader to design a turbocharger system for other cars.

The examples in this book are limited to four-stroke engines. Much of the information in this book is applicable to two-stroke engines, but there are other factors that need to be considered with two-stroke engines, which are not described herein. For example, any formulas relating engine airflow to engine speed would need to be multiplied by a factor of 2 when comparing two-stroke engines to four-stroke engines.

If your engine has a carburetor, I have you covered. Start by reading Appendix D.

© 2012 Robert G. Wagoner
All rights reserved.

Turbocharging Normally Aspirated Engines on a Budget

HP / Weight Ratio

Dividing the horsepower of a car by its weight gives a good indication of how fast it can accelerate. The table below is an interesting list of HP/weight ratio of the project cars listed in this book (before and after turbocharging) as well as some other well-known cars as references. The cars and motorcycles in this table are ranked in order from fastest to slowest.

	Make	Model	Year	Liter	HP	Weight	Note / Boost Level	Engine	HP/weight
	Yamaha	YZF-R1	2005	1.00	180	529	motorcycle	I-4	0.3403
	McLaren	F1	1997	6.10	680	2704	normally aspirated	V-12	0.2515
	Ferrari	Enzo	2002	5.99	660	2766	normally aspirated	V-12	0.2386
	Saleen	S7	2001	7.00	550	2755	normally aspirated	V-8	0.1996
	Chevrolet	Corvette Z06	2007	7.01	505	3162	normally aspirated	LS7 V-8	0.1597
	Lamborghini	Murcielago	2004	6.19	580	3638	normally aspirated	V-12	0.1594
App D	Chevrolet	Corvette	2003	5.67	467	3118	7 psi boost	LS1 V-8	0.1498
	Dodge	Viper	2004	8.30	500	3385	normally aspirated	SRT-10	0.1477
	Chevrolet	Corvette Z06	2004	5.67	405	3118	normally aspirated	LS6 V-8	0.1299
	Porsche	911 Turbo	2004	3.60	415	3388	factory turbocharged	flat 6	0.1225
	Chevrolet	Camaro	1970	6.50	375	3313	normally aspirated	V-8	0.1132
App D	Chevrolet	Corvette	2003	5.67	350	3248	normally aspirated	LS1 V-8	0.1078
	Ford	Cobra SVT	2004	4.60	390	3664	supercharged	V-8	0.1064
	Chevrolet	Camaro	2001	5.67	320	3306	normally aspirated	LS1 V-8	0.0968
CH 6	Toyota	Supra	1995	3.00	306	3219	8 psi boost	2JZ-GE I-6	0.0951
CH 4	Toyota	MR2 Spyder	2001	1.79	207	2195	9 psi boost	1ZZ-FED I-4	0.0943
	Pontiac	GTO	2004	5.70	350	3725	normally aspirated	LS1 V-8	0.0940
	Acura	NSX	2004	3.20	290	3153	normally aspirated	C32B V-6	0.0920
	Chevrolet	Camaro	1971	6.50	300	3313	normally aspirated	V-8	0.0906
	Nissan	350Z	2004	3.50	287	3188	normally aspirated	VQ35DE V-6	0.0900
	Cadillac	XLR	2004	4.60	320	3650	normally aspirated	Northstar V-8	0.0877
CH 7	Mazda	Miata	1997	1.80	205	2350	10 psi boost	BP-ZE I-4	0.0872
	Ford	Mustang	2003	4.60	300	3466	normally aspirated	V-8	0.0866
	Honda	S2000	2004	2.16	240	2835	normally aspirated	F22C1 I-4	0.0847
	Ferrari	Dino GTS	1972	2.40	195	2381	normally aspirated	V-6	0.0819
CH 5	Dodge	Neon R/T	1999	2.00	192	2385	6 psi boost	DOHC	0.0805
CH 3	Pontiac	Fiero	1987	3.40	231	2900	11 psi boost	3.4 L V-6	0.0797
	Mazda	RX8	2004	1.30	238	3053	normally aspirated	Wankel	0.0780
	Audi	TT quatro	2004	3.20	250	3351	normally aspirated	V-6	0.0746
	Pontiac	Grand Prix	2004	3.80	260	3583	supercharged	V-6	0.0726
	Acura	RSX type S	2004	2.00	200	2778	normally aspirated	K20A2 I-4	0.0720
	Chrysler	PT Cruiser	2004	2.40	220	3101	factory turbocharged	2.4L HO	0.0709
	Chevrolet	Monte Carlo SS	2004	3.79	240	3448	supercharged	V-6	0.0696
CH 6	Toyota	Supra	1995	3.00	220	3219	normally aspirated	2JZ-GE	0.0683
CH 4	Toyota	MR2 Spyder	2001	1.79	138	2150	normally aspirated	1ZZ-FED	0.0642
CH 5	Dodge	Neon R/T	1999	2.00	150	2385	normally aspirated	DOHC	0.0629
Ch 7	Mazda	Miata	1997	1.80	133	2350	normally aspirated	BP-ZE	0.0566
	Honda	Civic EX	2003	1.70	127	2474	normally aspirated	V-Tech	0.0513
CH 3	Pontiac	Fiero	1986	2.80	135	2850	normally aspirated	L44 (V-6)	0.0474
	Volkswagen	Golf	1997	2.00	115	2540	normally aspirated	I-4	0.0453
	Chevrolet	Camaro	1975	5.70	155	3723	normally aspirated	V-8	0.0416

© 2012 Robert G. Wagoner
All rights reserved.

Turbocharging Normally Aspirated Engines on a Budget

Supercharger and Turbocharger Overview

Power comes from burning fuel, but the amount of fuel an engine can burn is limited by the amount of air in the combustion chambers. It is easy to add fuel, and the difficult part is adding more air. Forced induction systems compress air to force more air into the engine's combustion chambers than would flow in a normally aspirated engine. More air in the cylinder allows an engine to burn more fuel, making more torque at any particular RPM, resulting in more horsepower without increasing RPM. Since the mechanical force increases with the square of the speed, increasing horsepower by turbocharging is less likely to cause engine damage than increasing horsepower by increasing RPM.

In this book the term turbocharging will refer to a system of forced induction driven by engine exhaust gas using a turbine to turn the compressor. A turbocharger is a compressor driven by exhaust gasses from the engine flowing through the turbocharger's turbine. Because a turbocharger is not a positive displacement pump, it works best at high RPM, and it does not produce much boost at low RPM, similar to a centrifugal supercharger. Typical operating speeds of a turbocharger are 80,000 to 140,000 rpm.

In this book the term supercharging will refer to a system of forced induction driven directly by the engine's crankshaft via a belt or chain to turn the compressor. A supercharger is a compressor driven by a pulley from the crankshaft. The majority of superchargers can be grouped into three categories based on style of compressor impellers: a centrifugal-style impeller, twin rotating screws, or counter-rotating rotors (roots-type supercharger). The latter two types are positive displacement pumps, which can create significant boost pressure at any RPM. The centrifugal-style is more like a turbocharger in that it pumps air best at high RPM and it does not produce much boost at low RPM. Typical operating speeds for the positive displacement types are 10,000 rpm to 15,000 rpm. Typical operating speeds for centrifugal superchargers are 20,000 rpm to 30,000 rpm.

For those interested, Appendix H provides information compares superchargers and turbochargers in areas of cost, efficiency, torque, and power.

Definitions and Terminology

Depending upon the source, some terms are used to mean different things. To avoid confusion, I will explain my use of the terminology in this book.

A/R Ratio is the area divided by the radius of the compressor or turbine housing. Turbine A/R ratio is very important because a large A/R ratio on the turbine is less restrictive at higher rpm allowing higher peak HP, whereas a smaller A/R ratio on the turbine will spool the turbo faster to provide more boost at lower RPM.

Absolute Pressure or ***Absolute Barometric Pressure*** is a measure of the air pressure compared to a vacuum. In this book the units will be "PSIa" to distinguish this from relative pressure. Absolute pressure of air is about 14.7 PSIa at sea level. See ***relative pressure*** for a comparison.

Air Filter or ***Air Cleaner*** is a device that removes the dust, dirt, and moisture out of the air to make the engine and turbocharger last longer. The air cleaner also reduces intake air noise. The air cleaners described in this book may be of two different types: dry types with replaceable filter elements and oil bath types, which can be cleaned and reused. The symbol used in this book for an air filter is shown at the right.

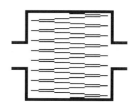

© 2012 Robert G. Wagoner
All rights reserved.

Turbocharging Normally Aspirated Engines on a Budget

Air Temperature Rise is an undesirable consequence when a turbocharger compresses the air. It is based on inlet temperature, the amount of boost, and the compressor efficiency. The detailed calculations of this air temperature rise are shown in Appendix B.

Blower is a common slang term for a supercharger, but it has many other meanings, such as an electric fan, and consequently the term ***Blower*** will not be used in this book except here in the definitions section.

Boost will be defined herein as the relative pressure (above outside air pressure) measured in the intake manifold. See relative pressure for more information.

Brake Specific Fuel Consumption, (***BSFC***), is a measure of the engine's fuel consumption per horsepower. Simply put, when producing a certain horsepower, if extra fuel is injected, this is "rich". For gasoline powered engines, the range for BSFC is typically between 0.45 to 0.6, in pounds of fuel per hour per horsepower, where 0.45 is very lean and 0.6 is very rich. The BSCF value is used in calculations of fuel injector sizing. See page 22 for details.

Carbon monoxide (CO) is an exhaust emission resulting from incomplete combustion, which is created in higher volume when running rich than when running lean.

Compressor will refer to the compressor side of the turbocharger, the side that compresses the intake air and provides boost.

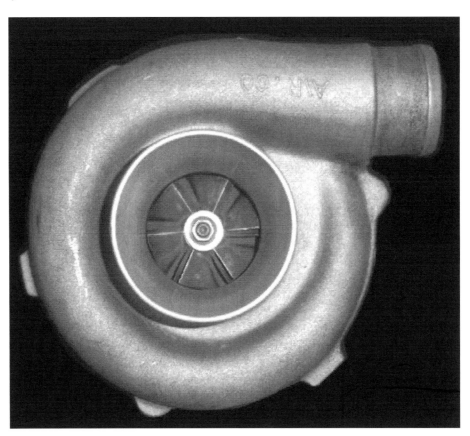

Cylinder Head or simply ***Head*** is the component in the induction system mounted directly to the engine block, which contains the valves to control the flow of fresh air in and exhaust out.

Density Ratio is the density of the turbocharger outlet air divided by to the density of the inlet air.

Turbocharging Normally Aspirated Engines on a Budget

Detonation is a violent uncontrolled burn of the air/fuel mixture, an explosion instead of normal burning. This is evident as a sound known as "spark knock" or "ping", and can damage an engine.

ECM (Engine Control Module), is the computer in the car that regulates fuel and spark timing.

Exhaust Back Pressure is the pressure measured at the outlet of the exhaust manifold, caused by restrictions in the exhaust. A turbocharger inserts a turbine into the exhaust flow causing significant exhaust back pressure, whereas a supercharger creates no additional exhaust back pressure.

Exhaust Manifold is the component in the induction system mounted directly to the head, which directs exhaust from the engine to the turbocharger. The exhaust manifold is commonly made of cast-iron. The symbol used in this book for an exhaust manifold is shown at the right.

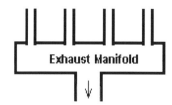

Forced induction is an induction system that includes either a supercharger or turbocharger.

Fuel Octane Rating is an indication of a fuel's sensitivity to knock , which is typically pressure-induced self-ignition. A higher octane fuel does not necessarily contain more specific heat. In other words, a higher octane fuel by itself will not naturally increase the engine power. The higher octane fuel will allow advancing the timing of an engine, or more boost or leaner AFR, resulting in higher cylinder pressure, and this does make more power.

Fuel Specific Heat Energy is the amount of energy released in burning a particular fuel. This factor is important to calculations of horsepower in this book.

Fuel Vaporization is a term used to describe the process of changing liquid fuel into a vapor so it burns properly in the engine.

Heat Capacity is the amount of heat required to raise the temperature of an object (or substance) by one degree Kelvin.

Hydrocarbons (HC) are a family of unwanted harmful exhaust emissions resulting from incomplete combustion, especially when running rich (white exhaust smoke).

Induction system will refer to the path air follows to get into the cylinders of the engine. It includes components such as the air filter, piping, turbocharger, intercooler, throttle body, and intake manifold.

Intake Manifold is the component in the induction system mounted directly to the head, which directs air from the air cleaner or turbocharger through plenum runners to each of the engine's cylinders. The intake manifold is commonly made of cast-iron or cast-aluminum alloy construction or a plastic composite material. The symbol used in this book for an intake manifold is shown at the right.

© 2012 Robert G. Wagoner
All rights reserved.

Turbocharging Normally Aspirated Engines on a Budget

Intercooler will refer to a heat exchanger, located between the turbocharger and engine at the output side of the turbocharger, which reduces the intake air temperature by absorbing some of the heat out of the compressed air. Intercoolers can either move the heat directly to the outside air, called air-to-air, or they can move the heat to a liquid cooling loop, called air-to-water. These are sometimes referred to as a "Charge-Air-Cooler".

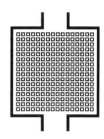

Knock see ***Spark Knock***.

Knock Sensor is a device that senses the sound associated with engine knock and sends a signal to the ECM, which is ordinarily used by the ECM to retard ignition timing.

Manifold Absolute Pressure (MAP) sensor is a device which measures the absolute pressure in the intake manifold and sends that information to the ECM, typically with a 0-5V signal.

Mass Airflow (MAF) sensor is a device which measures the amount of airflow in the intake piping and sends that information to the ECM, typically with a 0-5V signal.

Naturally Aspirated (NA) engine is an engine without forced induction.

Particulate Matter (PM) is a type of unwanted harmful exhaust emissions, made up of unburned particles of fuel and engine oil, which appears as black exhaust vapor and settles in and around the exhaust pipe as black soot. Like most other unwanted harmful exhaust emissions, this occurs more when running rich than when running lean, because when running rich all of the fuel cannot be burned.

Piggyback Fuel / Ignition Controller (FIC) is a separate ECM that works in conjunction with the factory ECM, which allows making changes to the fuel injector duty cycle and the ignition timing, based on boost pressure and other variables.

Positive Crankcase Ventilation (PCV) is a system that provides clean air into the crankcase, so the exhaust gas blow-by that slips past the piston rings does not collect in the crankcase. This extends engine oil life and has other positive environmental benefits.

Pounds per Square Inch (PSI) is the unit of measurement for pressure used in this book.

Pressure-induced self-ignition is a type of auto-ignition condition that occurs when the air/fuel mixture in the cylinder explodes (detonation) due to exceeding the self-ignition temperature of the air/fuel mixture, which makes the knock sound.

Pressure Ratio is the outlet absolute pressure of the turbocharger's compressor divided by the inlet pressure. The inlet pressure is usually slightly below atmospheric (14.7 PSI at sea level) due to pressure drops in the air filter and intake pipes.

Relative Pressure or ***Relative Barometric Pressure*** is a measure of the air pressure compared to the pressure of air at sea level. See ***absolute pressure*** for a comparison. Effectively relative pressure is equal to absolute pressure - 14.7 PSIa. Hence the absolute air pressure outside at sea level is 14.7 PSIa and the relative air pressure outside at sea level is 0 PSI.

Spark Knock is the pinging sound or knocking noise associated with detonation, caused by the shock of the explosion causing undesirable metal-to-metal contact inside the engine. This sounds like hitting metal with a hammer.

© 2012 Robert G. Wagoner
All rights reserved.

Turbocharging Normally Aspirated Engines on a Budget

Specific Heat Capacity or ***Specific Heat*** is the amount of heat required to raise the temperature of one gram of a substance by one degree Kelvin. This is related to amount of energy released in burning a particular fuel, and more information on this subject can be found in Appendix B.

Supercharger will refer to a pumping device mechanically driven directly from the engine crankshaft that increases the amount of air delivered to each of the engine's cylinders.

Supercharging will refer to a system of forced induction driven directly by the engine's crankshaft via a belt or chain to turn the compressor.

Turbine will refer to the turbine side of the turbocharger, the exhaust driven side, which drives the compressor.

Turbocharger will refer to a pumping device driven by exhaust gases from the engine's exhaust system that increases the amount of air delivered to each of the engine's cylinders. ***Turbo*** is another common name for a turbocharger, but it has many other meanings, so it will not be used in this book, except when turbo is part of a specific name of a car, and except for "Turbo Lag".

Turbocharging will refer to a system of forced induction driven by engine exhaust gas using a turbine to turn the compressor.

Turbo Lag is the delay between the moment the accelerator is depressed until the boost pressure in the intake manifold increases significantly.

Wastegate is a controlled valve that allows the exhaust gasses to bypass the turbine to limit the boost pressure.

Wide Open Throttle (WOT) is the term for an acceleration run with the gas pedal to the floor, used to test and collect data during tuning. I would never do this just for the fun of it. ☺

> **Warning**
> Modifications in this book may not be legal for street cars. Check local and state laws before making any changes to your car.
>
> Modifying your engine will void any factory warranties.
>
> The author does not recommend that anyone make any modifications to their car. The author is not responsible for any loss, damage, or injury caused by any modifications anyone makes to their car.

Expect Breakage
- *I have had multiple major failures when modifying engines to increase power.*
- *I have learned to start a project like this with an expectation of major damage, and I feel lucky if that doesn't happen.*
- *I have done the most damage when I decided at the end to bump up the boost more and more until I "find the limit".*

Reliability
- *Minimize the number of modifications – every change reduces reliability.*
- *Start with a good-running, low-mileage engine, clutch, and drivetrain.*
- *Ensure every modification is done carefully and correctly.*
- *Plan from the beginning on a conservative power increase – don't push it.*

> *This book should be used in conjunction with a factory service manual, or a repair manual for your specific vehicle from GM, Chilton, or Haynes, which will include safety procedures, engine rebuilding information, torque specifications, cleaning, etc.*

Turbocharging Normally Aspirated Engines on a Budget

2 Turbocharging: Making It Reliable

There are literally thousands of SAE papers describing analyses of modern engines, describing how the design engineers have analyzed every possible mechanical stress using Finite Element Analysis, every airflow twist and swirl, mechanical tolerances, oil pressure in every passage, cooling, temperatures, including statistical analysis and amount of margin on every one. Clearly there is no way for a person working on an engine in his garage, or even a group of a dozen experienced mechanics, to expect to understand every stress like the engine design engineers. Consequently, as we modify an engine, we must expect the possibility something might break. We find the limits the hard way, not by calculation, but by causing damage.

> Have you seen an engine explode during an NHRA race on TV? These engines are designed and assembled by the best engine builders. They break due to overstress in some component. The simple lesson: if they push it too hard, it will break.

That being said, it **is** possible to turbocharge a normally aspirated engine with high reliability... although never as reliable as an unmodified engine. This chapter describes some considerations for turbocharging a normally aspirated engine that will help make it as reliable as possible.

Designing for Reliability of the Turbocharger Itself

A turbocharger can be very reliable if a few simple precautions are observed. The turbine wheel is exposed to exhaust gas temperatures in the range of 1200°F to 1600°F, and it gets very hot. Cooling and lubrication are the most important things to keeping it happy. Supplying plenty of oil to the bearings is most important. As I install a turbocharger onto a normally aspirated engine, tapping into an oil supply with plenty of flow is absolutely necessary. Similarly, if the turbocharger is water cooled, tapping into a coolant source with plenty of flow is absolutely necessary. Also, supplying the turbocharger with clean oil is important, and adding the turbocharger onto a normally aspirated engine will cause the oil to degrade more rapidly than normal. Hence, it becomes more important on a turbocharged engine to perform oil and oil filter maintenance on schedule or ahead of schedule. For added protection, use synthetic oil. Finally, since the turbocharger takes some time to spin down, shutting off the engine immediately after revving the engine should be avoided. Instead allow the engine to idle for at least 30 seconds before shutting it off, to allow circulation of oil and coolant as the turbocharger spins down.

Designing for Reliability of the Entire System

- Reliability is higher with fewer components. To state that conversely, the more components in a system, the lower the reliability. If there are two ways of accomplishing the same task with equal benefits, choose the one that has fewer components.

- Electrical wiring splices and connectors are much more likely to fail than a solid wire. The fewer wires that need to be cut, the better. To make a wire connection, soldering is the most reliable, crimping is fairly reliable, and twisting wires together is not reliable.

- Every pipe/hose connection with a hose clamp is more likely to fail than a solid pipe. Hose clamps can come loose and leak. Similarly every plumbing fitting might leak. The fewer connections, the better.

Which one is more likely to leak?

Design for reliability.

© 2012 Robert G. Wagoner
All rights reserved.

Turbocharging Normally Aspirated Engines on a Budget

Designing for Reliability of the Engine: Detonation and Spark Knock

I will jump right into the middle of this one to bring out the terminology. Detonation causes spark knock. Detonation can be caused by pre-ignition. Pre-ignition, detonation, and spark knock are not the same. Let's start with definitions, to clear up some common misuse of terminology.

- *Pre-ignition* is basically the act of igniting the air/fuel mixture too early.
- *Detonation* is a violent uncontrolled burn of the air/fuel mixture, an abnormal type of combustion in an engine where a portion of the air/fuel mixture spontaneously self-ignites and burns much more rapidly than normal, i.e. an explosion instead of normal burning.
- *Spark knock* is the pinging sound or knocking noise associated with detonation, caused by the shock of the explosion making the piston to slap against the inside wall of the cylinder.

Normal combustion is initiated by a spark with the flame front spreading smoothly from the spark plug and traveling across the combustion chamber. After the spark ignites the air/fuel mixture, the cylinder pressure starts to increase as the fuel burns. As the cylinder pressure increases, so does the temperature. If the ignition timing is too advanced, the cylinder pressure and temperature can increase to a temperature above the fuel's self-ignition temperature, at which point detonation occurs. Chapter 3 includes a photograph of a piston we damaged by detonation from too much boost. Detonation can also damage rings, spark plugs, valves, head gaskets and rod bearings. Detonation can be caused by too much compression, carbon deposits in the combustion chambers, ignition timing too advanced, overheated air from too much boost, a lean air/fuel mixture, low octane fuel, and other factors. Any time I hear spark knock, I let off the throttle immediately to avoid damaging the engine, and make changes to add fuel (make AFR richer) and retard the timing as much as necessary to eliminate the spark knock.

Reducing Compression Ratio to Avoid Detonation

For the purpose of turbocharging an engine, it helps to reduce the compression ratio, which effectively allows cramming more air/fuel into the cylinder before having trouble with detonation. See the previous section on detonation and spark knock for more details. There are various ways to reduce compression ratio, such as changing pistons, changing heads, grinding the heads, or using thicker head gaskets. Reducing compression ratio is not as important for lower boost applications (such as 7 PSI boost) as it is for higher boost applications (such as 15 PSI boost). As an example of calculating compression ratio, below are the calculations performed for using 0.070" thick head spacer on a 3.4L V-6 engine to reduce the compression ratio.

The original 3.4L engine has bore = 3.62", and stroke = 3.31" and 9.0:1 compression ratio.

Original compressed volume = $\pi * (3.62"/2)^2 * 3.31 / 9.0 = 3.7852$ cubic inches

Additional volume = $\pi * (3.62"/2)^2 * 0.075" = 0.7719$ cubic inches

New compression ratio = $[\pi * (3.62"/2)^2 * 3.31] / (3.7852 + 0.7719) = 7.5 : 1$

Colder Plugs to Avoid Detonation

The temperature of the spark plug tip is important. It must be hot enough to keep itself clean, but it must be kept below the ignition temperature of the air/fuel mixture. If the spark plug tip gets too hot, it will cause "pre-ignition" to occur. In this particular case, ignition timing cannot correct this problem, since the fuel is being ignited by the hot tip of the spark plug instead of by the spark itself. See the section above on detonation and spark knock for more details. The temperature of the burning fuel is much higher than the temperature of the spark plug tip. This heat flows from the tip

© 2012 Robert G. Wagoner
All rights reserved.

of the spark plug center electrode through the ceramic insulator, and eventually through the head into the water cooling system. By design of the length of the tip and shape of the ceramic, the temperature of the tip can be controlled.

As the engine is modified to add power, more heat is produced in each cylinder, along with higher cylinder temperatures and pressures. This causes the spark plug tip to run hotter. A very competent spark plug manufacturer, NGK, recommends changing to one range colder plug for every 75 HP to 100 HP added to an engine. When we add a turbocharger, we are in this range, and a colder plug may be important.

The heat range numbers used by spark plug manufacturers are not the same. Each manufacturer has a different rating system. In fact, they do not even all go the same direction. Domestic manufacturers such as Autolite and Champion use a system where hotter plugs have higher numbers. The Japanese manufacturers such as Denso and NGK use a system where colder plugs have higher numbers.

It is possible to "read" the spark plugs. Pictures can be found in many automotive manuals. The ceramic insulator around the inner electrode should be grayish tan to white colored. If it is yellowish colored or blistered, the plug is running too hot, and a colder plug should be used. If they look dirty or dark brown they may be running too cold, and a hotter plug may be needed. However, it is rare to need a hotter plug than the one recommended by the manufacturer, so if they look dirty or dark brown this is more likely to be caused by running rich or burning oil. When selecting spark plug heat range, it is better to end up with a plug that is too cold than too hot. If the plug heat range is too cold the result will be a fouled spark plug, but if the plug heat range is too hot the result could be severe engine damage from pre-ignition.

Adding an Intercooler to Avoid Detonation

An intercooler located between the turbocharger compressor outlet and the engine intake manifold will cool the compressed air, providing more horsepower and less chance of detonation. For applications with low boost, such as below 8 PSI of boost, the turbocharger may be fed directly into the engine without an intercooler. To increase the boost above that level, it becomes more important to add an intercooler. In some cases it is very difficult to find the space to add an intercooler. When an intercooler cannot be added due to available space, this will limit the total boost to a lower level, resulting in less power than could be achieved with an intercooler. In keeping with the philosophy of adding components that provide significant performance gain for relatively low cost, an intercooler is a good component to add. Hence most of the examples in this book will include an intercooler. Much more information on intercoolers can be found in the next chapter describing the Excel spreadsheet for turbocharger compressor sizing.

Intercooler efficiency is based on the temperature of the inlet air to the intercooler compared to the temperature of the outlet air. These temperature measurements will be different based on many factors, particularly the external cooling airflow through the intercooler. Most certainly when an intercooler manufacturer measures intercooler efficiency, they will provide a strong source of fast moving, fresh air for the outside cooling to make the measurement. Consequently any intercooler efficiency number found in literature will not be exactly the same value when used on a particular vehicle. When an intercooler is mounted in an automobile, it needs to be located so it has a source of fast moving cool, fresh air to have the highest possible efficiency. For example, if front mounted, it NEEDS to be placed in front on the radiator, not behind it. I want to point out the intercooler airflow in a dynamometer test where the car is standing still with fans blowing at the engine is not the same as the airflow when the car is moving during a road test.

Turbocharging Normally Aspirated Engines on a Budget

An intercooler has some restriction to the airflow, so the pressure at the compressor outlet will be higher than the pressure at the intake manifold inlet if there is an intercooler. If there is an intercooler, its pressure drop must be subtracted from the total pressure to get the pressure at the engine intake manifold. As more air flow is forced through an intercooler, the pressure drop increases. Pressure drop is a function of the air flow, and pressure drop can be scaled by the ratio of the actual airflow compared to the rated air flow. For example, at half the airflow, the pressure drop is cut in half. Obviously everything should be as efficient as possible, but at these low boost levels, slightly higher pressure drop in an intercooler can be compensated by higher boost pressure at the turbocharger outlet. In other words, the wastegate could be adjusted to a higher pressure to compensate for intercooler pressure drop.

For the best performance, it is important to have the largest possible intercooler with the lowest pressure drop. When selecting an intercooler, the available space is the ultimate limitation. In general pick the largest intercooler that will fit. Intercooler volume is a good figure of merit for a intercooler: the larger the better. If various intercoolers will fit, chose an intercooler with more frontal area and less thickness, because these will tend to be more efficient and have less pressure drop.

Some intercoolers may be supplied with the data needed, but most are not. It is much more convenient to purchase an intercooler that includes data. The example values entered in the spreadsheet above are 1 PSI drop and 71% efficiency, which are the values we measured using a stock Volvo intercooler mounted in the trunk of a Fiero with its own dedicated cooling fan. Many smaller stock intercoolers have higher pressure drop and lower efficiency than this example. Most aftermarket intercoolers with thicker cores and larger pipes will have lower pressure drop at slightly higher efficiency than a small stock intercooler. Below are comparisons of these parameters for some intercoolers I have either measured or found in other literature.

Intercooler Type & Dimensions (HxWxD)	Pressure Drop	Efficiency
Stock 300 ZX (6.5" x 6.8" x 2.2")	3 PSI at 160 CFM	53%
Stock 911 Porsche (9.5" x 16" x 3.5")	0.5 PSI at 430 CFM	51%
Stock Volvo 240 (17" x 18.5" x 1.1")	1 PSI at 330 CFM	71% (with dedicated cooling fan)
Stock BMW N54 (5" x 26" x 3")	2 PSI at 300 CFM	82% (see graph below)
Stock Turbo Regal (9.8" x 12.2" x 2.8")	3.5 PSI at 350 CFM	68%
Stock Subaru WRX Top Mount	0.7 PSI at 200 CFM	71%
Stock Subaru WRX Top Mount	1.6 PSI at 420 CFM	63%
Aftermarket HKS (10.2" x 7.8" x 2.5")	1.4 PSI at 330 CFM	68%
Aftermarket Stillen (9.4" x 8.4" x 3.5")	1 PSI at 330 CFM	66%

Intercooler efficiency is a function of airflow, as shown in the graph on the next page. This makes it even more difficult to use a single number provided by a manufacturer, because they don't provide the efficiency curve. The graph below is for a stock intercooler from a BMW E90 335i N54 3.0L twin turbo. Many people are upgrading these stock intercoolers to larger aftermarket intercoolers, because these cars can get a significant performance increase just by switching to a better intercooler. Consequently the stock BMW intercoolers are readily available on the used market for a low price, and they work well in most of the applications in this book.

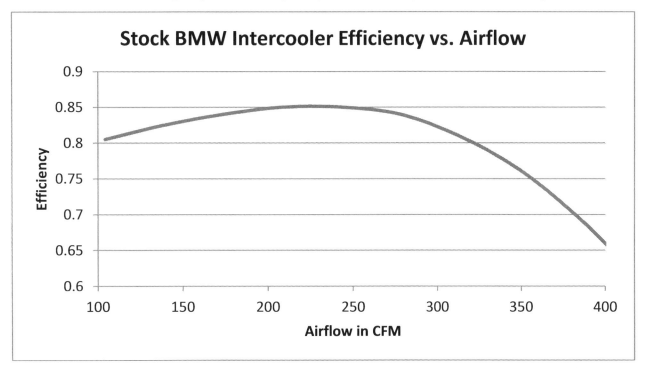

The intercooler pressure drop and efficiency can be measured after the intercooler is installed with pressure and temperature gauges, by measuring the inlet and outlet values at WOT under load at high RPM. I have been able to take such measurements in a higher gear on a highway during a short burst of acceleration going uphill at roughly the speed limit.

- To measure intercooler efficiency, use the MAF sensor as the indication of airflow, along with the outside air temperature, inlet air temperature, and outlet air temperature. When measuring temperatures, it is important to measure the air temperature inside the piping. Some sources indicate it is OK to measure the temperature of the pipes themselves, but I have found measuring the air temperature is more accurate.

- The most accurate way to measure pressure drop is to use a differential pressure gauge, which measures the pressure drop directly, instead of using two separate pressure gauges and subtracting the readings. A single differential pressure gauge avoids the errors in calibration between two different boost gauges

Sizing Fuel injectors for Reliability

Running lean can damage the engine quickly, and to make a reliable turbocharged engine, the fuel system needs to be considered carefully. As more air is fed into the engine, more fuel is needed to match. This could be done by increasing the fuel pressure, or increasing the fuel injector size, or a combination of both. Such changes need to be coordinated with the entire fuel system, considering the limitation of the fuel pump pressure vs. flow curve. It may not be possible to increase the fuel pressure enough to get the necessary flow through the original fuel injectors, in which case it will be necessary to change the fuel injectors. In fact increasing the fuel pressure is only useful where a small change is needed in total fuel. If a lot more fuel is needed, changing the fuel injectors will be necessary.

Before calculating fuel injector size, it is important to understand the impact of fuel pressure. The reason this is important is because the fuel pressure affects fuel injector flow rate. Specifically, fuel injector flow rate is proportional to the square root of the fuel pressure ratio. The fuel injectors have a rated flow rate that is usually measured in LB/hr. For many fuel injectors, this flow rate of the fuel injector is based on 43.5 PSI of pressure drop across the fuel injector, with the fuel injector duty cycle being 100%. If the fuel pressure in a system is set at a different pressure, for example 50 PSI, the actual fuel flow will be adjusted by the square root of that ratio, by a factor of $(50/43.5)^{0.5} = 1.08$ in this example.

The standard formula for calculating fuel injector size is shown below.

$$\text{Fuel injector size (LBS/hr)} = \frac{\text{Maximum Horsepower} \times \text{BSFC}}{\text{number of fuel injectors} \times \text{Maximum fuel injector duty cycle}}$$

where BSFC is Brake Specific Fuel Consumption, the number of pounds of fuel per HP per hour.
{ See the next section for more information about BSFC }

Turning the formula above around to calculate maximum HP results in:

$$\text{Maximum HP} = \text{Fuel injector duty cycle} \times \text{Fuel injector size} \times \text{Number of fuel injectors} / \text{BSFC}$$

Using the Fiero as an example, there are six fuel injectors, each rated at 19LB/hr. If the fuel pressure was 43.5 PSI (at 100% duty cycle and BSFC = 0.6 to be very rich) results in the upper limit of maximum HP for these fuel injectors, as calculated below.

$$\text{Maximum Horsepower} = 1.0 \times 19 \times 6 / 0.6 = 190 \text{ HP}$$

BSFC, Brake Specific Fuel Consumption

BSFC is the engine's fuel consumption in pounds of fuel / hour / horsepower. The units for this variable are purposely coordinated with fuel injectors for ease of calculating fuel injector size. For gasoline powered engines, the range for BSFC is 0.45 to 0.6, where 0.45 is very lean and 0.6 is very rich. When an engine is at WOT (Wide Open Throttle) the air/fuel ratio is set rich to avoid any potential problems from high exhaust gas temperature. Turbocharged or nitrous oxide injected engines should run a richer air/fuel ratio than normally aspirated engines. For estimates used in calculations of fuel injector sizing, set BSFC = 0.5 for normally aspirated engines, and BSFC = 0.6 for turbocharged or nitrous oxide injected engines.

Blow-Off Valve (BOV)

In terms of reliability, adding a Blow-Off Valve (BOV) is generally considered a reliability improvement, as a means to avoid compressor surge. However, adding unnecessary components hurts reliability, so the previous statement is not true if compressor surge is not a problem. Compressor surge is a phenomenon where the air pressure at the compressor outlet is higher than the compressor can maintain, resulting in reversal of airflow through the compressor. This occurs when the throttle is closed quickly, such as during gear changes on a manual transmission car. The BOV is controlled by the manifold vacuum, so that when the driver lets off the throttle suddenly, the BOV opens to release the excess pressure from building up at the outlet of the turbocharger. With very low levels of boost, it is possible to build a system without a blow-off valve. However, even at 7 PSI of boost and above, I prefer to install a BOV. Since the BOV opens the intake to another source of air, an extra air filter should be added to the outlet of the BOV. Venting a BOV to open air without a filter will allow dirt to get into the engine, and this will cause the engine to wear out much faster.

In systems with a MAF sensor, the BOV should be routed in a way to allow the MAF sensor to sense total airflow to the engine at all times (when the BOV is open and closed). This may require

Turbocharging Normally Aspirated Engines on a Budget

the BOV outlet to route back into the intake piping at the turbocharger compressor inlet, in cases where the MAF sensor is located at the turbocharger compressor inlet. When the BOV is open, if it allows air to get into the engine that is not sensed by the MAF, this will result in the Air-to-Fuel ratio being incorrect, because Air-to-Fuel Ratio is controlled by the MAF sensor input to the ECM.

Wastegate

It is obviously very important to control the boost pressure to create a reliable system. A wastegate is a pressure controlled valve that allows the exhaust gasses to bypass the turbine. While some wastegates are controlled with an electrical solenoid, all of the wastegates described in this book are operated directly with boost pressure. They have a spring-loaded diaphragm that holds it closed, with boost pressure on the opposite side of the diaphragm, and when the boost pressure exceeds the spring force, the wastegate opens. A wastegate actuator for an internal wastegate is shown at the right. Sometimes a system with an internal wastegate has "boost creep", which is a situation where the boost pressure continues to rise as the rpm increases, resulting from the small size of the orifice in the internal wastegate.

Wastegate Actuator

If a system has "boost creep", it may become necessary to use a large external wastegate, which has a much larger orifice to bypass the turbocharger than internal wastegates, and consequently they do a better job of regulating the boost pressure. An example of an external wastegate is shown at the left. These require more exhaust plumbing than internal wastegates. An external wastegate is connected between the exhaust manifold and the turbocharger downpipe to allow the exhaust to bypass the turbocharger turbine when the external wastegate opens.

For the fastest response and best control stability, the boost pressure feed to the wastegate should come directly from the turbocharger outlet, in order to minimize delay time and provide the most stable boost pressure. This is contrary to some bad advice found on the internet, which may recommend sensing the boost pressure after the intercooler or somewhere else in the system as far away as the intake manifold. → A system with poorly regulated or fluctuating boost pressure is difficult to tune because of the unpredictability of the boost pressure at any given time. Hence this can impact reliability as well as performance.

The easiest way to adjust the operating pressure to a higher value is by adding a boost controller. An adjustable boost pressure regulator is shown at the right. This is a simple device that uses a ball with a spring and a screw to control the pressure. The wastegate actuator supplied with a turbocharger sets the boost at some nominal pressure, such as 8 PSI. To increase the boost beyond this, one of these adjustable boost pressure regulators can be added, which is connected between the turbocharger outlet and the wastegate actuator.

Adjustable Boost Controller

Turbocharging Normally Aspirated Engines on a Budget

Engine Mounts

Simply put, more torque may require stronger engine mounts. This impacts reliability because if the engine moves too much, it can break the exhaust pipes or intake plumbing. I suggest deciding on this after the turbocharger is installed and working properly. Most engine mounts will have enough strength to tolerate the extra torque that will be added for these low boost applications. Old/broken mounts may need to be simply replaced with standard motor mounts, new from the manufacturer. If stronger mounts are required, polyurethane mounts are available for most cars. I avoid these except when absolutely necessary because polyurethane mounts will transmit more vibration from the engine to the car.

Clutch

In general, when turbocharging a NA engine, most stock clutches are designed with enough margin to tolerate an increase in torque from a boost pressure of at least 7 PSI. Of course, increasing engine torque may require a stronger clutch. I have sometimes arbitrarily replaced a stock clutch with a high performance clutch in cases where I added a lot of boost pressure and when I planned to use that car primarily for track racing use. Replacing the stock clutch is ordinarily not necessary when adding a turbocharger with relatively low boost.

Tuning: Reprogramming the ECM

In terms of engine reliability, it is absolutely necessary to ensure the Air-to-Fuel Ratio is correct after adding a turbocharger. In some cases the turbocharged system may utilize a special rising rate fuel pressure regulator FMU (Fuel Management Unit) to control the air/fuel ratio by adjusting the fuel pressure {see the Dodge Neon chapter for an example}, but more often the air/fuel ratio is adjusted by changing the fuel injector duration by the ECM. It is also necessary to ensure the ignition timing is retarded as necessary to avoid detonation. The process of changing the Air-to-Fuel Ratio and ignition timing is called tuning.

When I am able to reprogram an ECM (Engine Control Module), I prefer to do that instead of changing the ECM to an aftermarket unit or adding an aftermarket piggyback FMU (Fuel Management Unit) to the existing ECM. Ordinarily reprogramming an ECM is less expensive and more reliable than the other options. In some cases the fuel systems and ECM are designed from the factory with enough margin to allow the addition of a turbocharger with absolutely no change to the ECM {see the Toyota Supra chapter for an example}.

Measuring Air-to-Fuel Ratio (AFR) is necessary to make adjustments to control it. In some cases the stock fuel system includes a wideband oxygen sensor that feeds signals into the ECM, and the Air-to-Fuel Ratio can be monitored directly from the ECM. Otherwise an aftermarket air/fuel ratio sensing system can be used during the tuning process, such as the UEGO (Universal Exhaust Gas Oxygen) sensor/controller manufactured by AEM. Similarly a knock sensor is an important sensor for tuning, and these are included in most modern fuel systems, but if not, an external knock sensor should be added during the tuning process.

> **Very Important Tuning Tip for Reliability**
>
> If you ever hear a ping (knock) back off the throttle immediately. See section on Detonation for more information.

Turbocharging Normally Aspirated Engines on a Budget

Camshafts for Turbocharged Engines

I prefer to change as little as possible, based on cost and reliability. My goal is to get the most "bang for the buck", which requires carefully selecting the components that are changed. Consequently, I use the stock camshaft, which is usually designed for more torque at lower RPM instead of the highest horsepower at high RPM, known as a "mild" grind. With moderate levels of boost the stock camshaft for the normally aspirated engine works just fine for a turbocharged engine. Adding a turbocharger to such an engine will result in an engine with plenty of torque at lower RPM, and still a significant power increase too. For readers interested in selecting a different camshaft, Appendix G discusses camshafts specifically for turbocharged applications.

Start with a Solid Engine and Drivetrain

This is probably obvious, so I will not discuss it at length. Simply put, for an engine to be reliable after turbocharging it, it must start with a good-running, low-mileage engine with a solid bottom end. Conversely, if the original NA engine doesn't run right, or if it is worn out already, or if has some strange noise, turbocharging it is a BAD idea. If the car has a check-engine light, fix it before turbocharging it. Similarly, the transmission, clutch, and entire drivetrain needs to be solid initially.

> **_Warning_**
> *Modifications in this book may not be legal for street cars. Check local and state laws before making any changes to your car.*
>
> *Modifying your engine will void any factory warranties.*
>
> *The author does not recommend that anyone make any modifications to their car. The author is not responsible for any loss, damage, or injury caused by any modifications anyone makes to their car.*

Expect Breakage
- *I have had multiple major failures when modifying engines to increase power.*
- *I have learned to start a project like this with an expectation of major damage, and I feel lucky if that doesn't happen.*
- *I have done the most damage when I decided at the end to bump up the boost more and more until I "find the limit".*

Reliability
- *Minimize the number of modifications – every change reduces reliability.*
- *Start with a good-running, low-mileage engine, clutch, and drivetrain.*
- *Ensure every modification is done carefully and correctly.*
- *Plan from the beginning on a conservative power increase – don't push it.*

> *This book should be used in conjunction with a factory service manual, or a repair manual for your specific vehicle from GM, Chilton, or Haynes, which will include safety procedures, engine rebuilding information, torque specifications, cleaning, etc.*

3. Turbocharging a 3.4L V6 Engine in a 1986 Pontiac Fiero

This is the first example of turbocharging a normally aspirated engine for less than $1000. This chapter will focus more on the hardware, and the next chapter will go into more tuning details.

1986 Pontiac Fiero

Turbocharged 3.4L Engine without Intercooler

Turbocharging Normally Aspirated Engines on a Budget

Turbocharger Selection

My previous book, "High Performance Fieros", provides details of an engine swap from the original 2.8L to the 3.4L. I will not go into that detail again here. That book also provides details of some experiments that were unsuccessful, like pushing the boost up until the engine failed. Some of that detail is omitted herein, to focus this book on a configuration with lower boost that is reliable.

Putting a 3.4L engine into a Fiero was certainly an improvement in performance as compared to the 2.8L engine. It was fine for a while, but eventually we decided we wanted more power. Since we had recently put in the 3.4L engine, and the engine was solid, we didn't want to make another engine swap. Instead, we decided the best thing to do was to turbocharge it.

> **The Progression from 2.8L to 3.4L**
>
> GM introduced the 60-degree V6 engine in 1980. The 2.8L version came first, and was the only size available through 1989. Its bore and stroke are 3.50" and 2.99" respectively. In 1990 the 3.1L version was introduced, with the same bore as the 2.8L, but a longer stroke of 3.31". The next step up in size came in 1991, with the 3.4L engine, this time by increasing the bore to 3.62", while keeping the 3.31" stroke of the 3.1L engine.

Our turbocharger selection was based on desired boost and expected HP. We started small, with an initial goal of approximately 200 HP. By calculation this took about 7-8 PSI boost. This seemed like the easy first step up, because no intercooler was necessary, keeping the plumbing simple. The Garrett T3/T03-"45" turbocharger and 19 LB/hr fuel injectors seemed like a good choice at this level. This setup is shown in the photograph on the previous page. After we got this working, we measured the power and found it was 187 HP with 7 PSI boost pressure, limited by the 19 LB/hr fuel injectors we were using at that time. This small turbocharger without an intercooler would be a good choice for this power level, operating this small turbocharger well within its capability.

We changed to 24LB/hr fuel injectors, added an intercooler and a BOV (Blow-Off Valve) and increased the boost until we had pushed this little turbocharger to its airflow limit. We found that it could put out 12 PSI boost up to about 4000 RPM, but dropped to about 10 PSI at 6000 RPM. At this point we measured 235 HP, but we decided we wanted more power. (Are you sensing a trend here?)

Now starting at 235 HP, it takes a bigger step to notice a difference. We decided our next goal was 300+ HP. By calculation this took about 20+ PSI boost. The Garrett T3/T04-50 turbocharger seemed like a good choice at this level. We installed this turbocharger and 30LB/hr fuel injectors, and with some difficulty got it working. We were making runs and slowly increasing the boost between runs. We finally got it up to about 21 PSI boost before the 3.4L engine had its first major failure! The first major breakage was a piston. Unfortunately we didn't get a chance to measure the power with 21 PSI boost. Above 18 PSI boost there was spark knock that we couldn't eliminate even with extremely retarded ignition timing and a very rich air/fuel mixture. And when we got to 21 PSI boost, the piston broke the first time we stomped on it. Aluminum chunks flying around in a cylinder cause major damage. We rebuilt the engine, replacing the damaged piston, honing the cylinder, and replacing the head due to a bent valve. After many more tests and adjustments, including more failures and another rebuild, we concluded a for a reliable turbocharged 3.4L engine with an intercooler the boost level should be limited to about 11 PSI. The turbocharger finally selected for this engine is a Garrett T3/T04-50.

Damaged Piston.

© 2012 Robert G. Wagoner
All rights reserved.

Turbocharging Normally Aspirated Engines on a Budget

Using the Excel Spreadsheet for Turbocharger Compressor Sizing

I like to start by using the Excel spreadsheet for turbocharger compressor sizing in Appendix A to pick the turbocharger and to get an estimate of the expected horsepower. This Excel spreadsheet can be downloaded from **WagonerEngineering.com**. Below is the Excel spreadsheet with values for the recommended turbocharger at the recommended boost level described in this chapter. The operating point is plotted on a compressor curve graph below the Excel spreadsheet.

	INPUTS	Source: "High Performance Fieros" by Robert Greg Wagoner
Outside Air Ambient Temp in deg F	60	
Altitude above Sea Level in ft	0	
Engine Volume in Liters	3.40	
Engine Compression Ratio	7.50	
Engine RPM for Maximum HP	5000	
Engine Volumetric Efficiency VE	0.700	Typical range is 0.70 to 0.85.
Engine Air to Fuel Ratio AFR	13.00	Should be rich. Enter 13.0 to 13.5.
Turbocharger Boost Pressure in psi (Gauge)	11.00	Set this to 0 if there is no turbocharger.
Turbocharger Compressor Efficiency CE	0.78	Find CE on the compressor map.
Intercooler Efficiency	0.71	Set this to 0 if there is no intercooler.
Intercooler Pressure Drop in psi	1.00	Set this to 0 if there is no intercooler.
X & Y VALUES FOR CE GRAPH		*formulas*
X = Air Flow in LBS / minute	25.5	B21*B22
Y = Turbocharger Boost Pressure Ratio PR	1.75	(B9+B19)/B19
OUTPUTS		
Outside Air Pressure in psia (Absolute)	14.7	14.7-B3*0.000494258
Boost Pressure Ratio after Intercooler	1.68	(B19+B9-B12)/B19
Air Flow in cubic ft / minute	333.7	0.0177516*B4*B6*B7*B10*B20/(B10+(1-B11)*(B16^0.283-1))
Air Density in LBS / cubic ft	0.076	2.7*B19/(B2+460)
Engine HP	231.2	298.2*B21*B22/B8*(1-1/B5^0.35)*(1-ABS(14.7-B8)/14.7)^2
Engine Torque in Ft-Lb	242.9	16500/pi()/B6*B23
Turbocharger Outlet Temperature in deg F	174.2	(B2+460)*(B16^0.283-1)/B10+B2
Intercooler Outlet Temperature in deg F	93.1	B25-B11*(B25-B2)

Excel spreadsheet with parameters for a Garrett T04-"50" operated at 11 PSI with an intercooler.

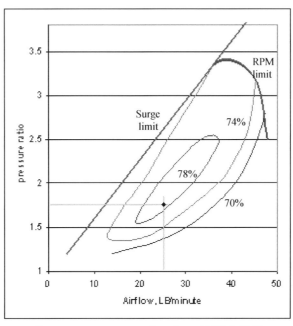

Compressor curves for Garrett T04-"50"

© 2012 Robert G. Wagoner
All rights reserved.

Turbocharging Normally Aspirated Engines on a Budget

Turbocharger System Block Diagram

Part1 of Chapter 4 provides an example to explain how to compare options and select the best system configuration. Without going into that detail here, below is the rationale and the Turbocharger System Block Diagram I selected or this engine.

Since the existing fuel control is based on a MAP sensor, to keep the cost as low as possible, it would be best to use the existing ECM and reprogram it. This ECM utilizes primarily the MAP sensor, exhaust oxygen sensor, and the throttle position sensor to adjust the Air-to-Fuel ratio and ignition timing. The MAP sensor is not designed for boost pressure, so it will need to be changed to a MAP sensor designed for an engine with boost. Besides that the fuel injector size will need to be increased. The stock fuel pump in the Fiero can support the required extra fuel flow needed up to the recommended 11 PSI boost and 231 HP, as described in detail below.

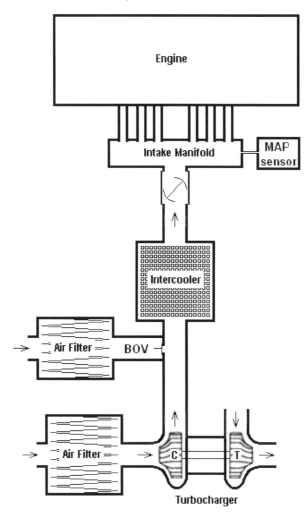

To turbocharge this 3.4L V6 engine in a 1986 Pontiac Fiero, the inlet air is pulled from a scoop in the rear quarter panel of the driver's side of the car where the air filter is located. From the air filter the air will feed into the turbocharger compressor inlet, then to the intercooler (located in the trunk), and finally into the throttle body, which is attached to the intake manifold, as shown below. A Blow-Off Valve (BOV) is located on the pipe at the inlet side of the intercooler, also located in the trunk. The BOV has its own small air filter attached directly to it, because when it is open it creates a source of air to the engine intake that needs to be filtered.

Turbocharging Normally Aspirated Engines on a Budget

Low cost parts and sources, new and used

To do these projects successfully at very low cost, using eBay is a must. In some cases I try to buy used parts, and in others I prefer new. Similarly in some cases I find the no-name brands work fine, and in other cases the well-known parts are worth the extra money. Here are some examples.

Intercooler: Since intercoolers have no moving parts, they don't wear out. As long as they are not physically damaged, which is obvious by a visual inspection, I prefer to buy a used intercooler from someone on eBay to save money. When I buy a new intercooler, the no-name parts are fine.

Intercooler piping: Generally I buy a universal intercooler piping kit. These kits usually include silicone connectors and hose clamps. Depending on whom you ask, some people like these, and others don't. I like them because they are very cost effective. They do, however, require more time to install then the prefabricated kits. It takes some type of saw to cut the intercooler piping. I find it is faster to use a body grinder with a metal cutting blade.

Exhaust manifold: When possible I prefer to use a cast iron exhaust manifold that can directly mount the turbocharger without any exhaust between them. They are usually less expensive and stronger than headers, and certainly less expensive than any custom exhaust pipe. I find these are most convenient, because the turbocharger can directly mount to it without an extra support bracket. This is a common practice on factory turbochargers such as the PT cruiser. All other examples in this book In this particular case (the Fiero) a special Y-pipe is necessary, and it is not possible to use a cast iron exhaust manifold that can directly mount the turbocharger.

Turbocharger itself: I will continue to give you the benefit of my experience, so that you can learn from my mistakes. Rebuilding a name-brand turbocharger sounds like a good idea to save money, but I have found rebuilt turbochargers to be unreliable. When a turbocharger is worn out enough that it needs rebuilt, the shaft is probably worn so much that it won't have the proper tight tolerances reliable is has been rebuilt. Without going into all the details, I have purchased a used Garrett T3 turbocharger and rebuilt it myself, and in another case I purchased one that was already rebuilt. Neither of them lasted long. In contrast I have purchased a number of the no-name turbochargers that worked very well. In conclusion, based on my experience, used or rebuilt name-brand turbochargers are not as reliable as a new no-name turbocharger. If you plan to use a rebuilt turbocharger, get one rebuilt by a professional that comes with a guarantee.

> Turbochargers need to be mounted so that the shaft is oriented horizontally, because their bearings are not designed for loading in a vertical orientation. It is not a good idea to place a flex pipe between the engine and the turbocharger because the high temperature and exhaust pressure can damage the flex pipe. The turbocharger needs to be mounted solidly to the engine. It can be supported directly with a "log" cast iron exhaust manifold. However, if exhaust pipe is used between the exhaust manifold and the turbocharger, steel support brackets must be added between the engine and the turbocharger to reduce mechanical stress on the exhaust pipe.

Initial Build with the "Small" Turbocharger

Turbocharging the Fiero turned out to be a bigger job than the 3.4L engine swap. The 3.4L engine swap took approximately 3 weekends. In comparison, turbocharging the Fiero took approximately 3 months. Everything goes on the engine from the top, so the hardware installation only took a few weekends, but the majority of the time was spent testing and tuning Air-to-Fuel Ratio (AFR) and adjusting timing curves in the Fiero ECM.

Turbocharging Normally Aspirated Engines on a Budget

The Turbocharger: We started with a Garrett T3 with a "45" size compressor, which is referred to as the "small" turbocharger in this chapter. More details about the Garrett T3 turbocharger described in more detail in Chapter 8, where it is used in a more appropriate application, a turbocharged 2.8L Fiero. We quickly pushed this Garrett T3/T03-45 turbocharger to its limit and moved up to a Garrett T3/T04-50 turbocharger, which is referred to as the "large" turbocharger in this chapter.

Turbocharger Location: The turbocharger is located in the position where the original Fiero air filter canister sat previously. It is supported from the bottom with a custom steel bracket mounted to the transaxle. This bracket is a simple steel strap, 1/8" x 1/2" with two bends at the ends to align it with the face of the turbocharger and the top of the transaxle. The turbocharger also receives some support from the "Y" pipe, but it should not actually apply much pressure on the "Y" pipe, because it can cause damage to the "Y" pipe.

Exhaust: The "Y" pipe that connects to the exhaust manifolds was cut and modified to feed directly into the turbocharger. This new "Y" pipe sits above the transaxle, as shown in the photograph below. The EGR valve has been removed and blocked. The turbocharger exhaust outlet is positioned downward. A custom exhaust down pipe was built to go between the turbocharger exhaust outlet and the standard Fiero exhaust. This custom exhaust pipe includes the lower flange and the oxygen sensor from the original Fiero "Y" pipe. The catalytic converter had been removed previously for better exhaust flow, and was left off when we turbocharged the Fiero.

New "Y" pipe.

Turbocharging Normally Aspirated Engines on a Budget

Mandrel Bent Exhaust Tubing: Improved Flow at Higher Cost

Mandrel bent pipes have less restriction to airflow than serrated bent pipes. Two examples of mandrel bent pipes are shown in the photograph below, a 1.75" diameter on the left, and a 3.00" diameter on the right. The mandrel bending process maintains a constant inside diameter through the bends. Even more importantly to reducing airflow restriction, mandrel bent pipes do not have the bumps and ridges that are found in serrated bent pipes. The equipment required to make mandrel bent tubing is more expensive than the equipment required to make serrated bends. For this reason, most muffler shops are not able to make mandrel bent pipes. This also explains why mandrel bent pipes are more expensive and harder to find than serrated bent pipes. Since mandrel bent pipes are more expensive than serrated bent pipes, local auto parts stores (which have to be competitive with other local auto parts stores) usually only have serrated bent pipes. Mandrel bent pipes can be found in speed shops where performance is more important than cost.

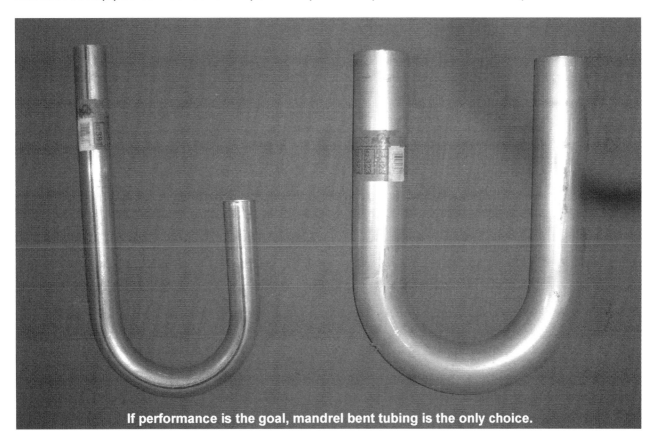

If performance is the goal, mandrel bent tubing is the only choice.

Connecting Two Exhaust Pipes Together

When connecting two pipes together to improve flow, cut the edges of the two pipes so that they meet together flush with no gap, and weld a bead on the outside with a MIG welder, so that the joint between the two pipes has a bump on the outside but not on the inside. If possible inspect the joint from the inside to make sure the inside is smooth. Depending upon a person's welding skill, especially if the welder is set too hot, sometimes there is a blob of weld on the inside that needs to be removed with a grinder. With practice it is possible to make these joints consistently without leaving any blobs on the inside. First practice on some scrap pieces to get the welder set up right before working on the actual parts to be used for the project. A MIG welder with thinner wire (0.030" diameter) and C25 gas (25% CO_2 and 75% Argon) is a good setup for welding thin steel like exhaust.

Oil system

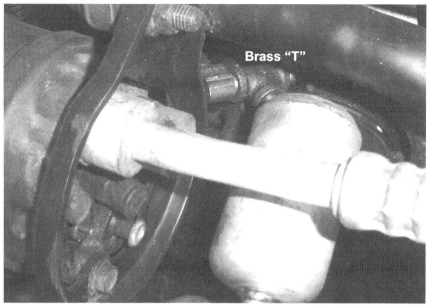

The turbocharger needs a good supply of oil. In order to keep the original location for the oil pressure sensor, a brass "T" fitting is added at the block to feed both the oil pressure sensor and the turbocharger. The hole for the oil pressure sensor tube is a different size on the 3.4L engine than it was on the 2.8L engine. This hole is directly beside the oil filter. For the 3.4L engine, the tube needs a brass adapter, 5/16" female for flared tube to 1/4" male pipe thread. Originally we put this together without air conditioning. This made the plumbing easy. Notice in the photograph above that the air conditioner compressor sets very close to these fittings. If the air conditioner is not used, as was the case for this Fiero originally, the exact positioning of these oil lines and fittings is not critical. However, when this engine was removed from this Fiero, the next Fiero that it was going into does have air conditioning. Adding the air conditioning caused changes from the list of fittings above. In the photograph above there is a tight 90° elbow that was added at the end of the brass "T" fitting to avoid the air conditioner and its lines. Oil return is gravity feed into the block through the hole for the camshaft position sensor. It is important to insure this oil return is large enough (1/2" diameter) to allow the oil to flow freely, and that it flows downhill the entire way. Also the oil return hose must be compatible with the oil. Do not use ordinary coolant line because it will fail quickly.

Intake and PCV

The entire intake plumbing and air filter from the original Fiero were removed and discarded. This includes the air filter canister. That area is now taken up by the turbocharger itself. A new intake and air filter setup was constructed simply with an aftermarket cone air filter and a straight pipe between this filter and the turbocharger inlet. By eliminating the original intake plumbing, the inlet for the PCV system was left unfiltered, so we added a small air filter directly at the crankcase air inlet, red in color, to match the valve covers.

Turbocharging Normally Aspirated Engines on a Budget

Manifold Absolute Pressure (MAP) sensors

The original GM MAP sensor (1 BAR), part number 2503679, is identified with the number 679 stamped into the plastic housing. Other 1 BAR GM MAP sensors have numbers 039, 460, or 883 stamped into the plastic housing. The 1 BAR MAP sensors are good for engines without a turbocharger, capable of measuring up to one atmosphere of absolute pressure, which is 14.7 PSI at sea level. Based on my measurements, the output voltage of the 1 BAR MAP sensor is shown in the graph at the right. Its output is linear, and can be plotted with the equation:

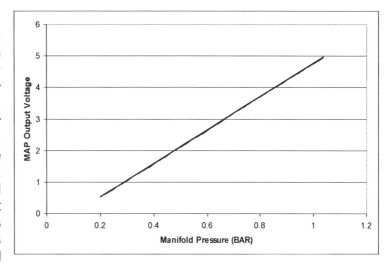

$$y = 5.2921x - 0.5474$$

2 BAR MAP sensor.

The 2 BAR GM MAP sensor (2 BAR), part number 16009886, is identified with number 886 stamped into the plastic housing, as shown in the photograph on the right. These can be found on 1984 to 1990 Pontiac Sunbird Turbo cars. Other two bar GM MAP sensors have numbers 012, 539, or 609 stamped into the plastic housing. The 2 BAR MAP sensors are capable of measuring up to 2 atmospheres of absolute pressure, good for engines with a turbocharger with up to 14 PSI of boost. We used a 2 BAR MAP sensor with the small turbocharger, and it ran and idled well (after much work reprogramming the Fiero ECM). This is the MAP sensor that should be used with the "recommended" 11 PSI of boost.

The 3 BAR GM MAP sensor, part number 16040749, is identified with the number 749 stamped into the plastic housing. These were used on the turbocharged 3.8L V6 engine in the 1989 Turbo Trans AM. The 3 BAR MAP sensors are capable of measuring up to 3 atmospheres of absolute pressure, good for engines with a turbocharger with up to approximately 28 PSI of boost. We used this 3 BAR MAP with our large turbocharger, but we were never able to get the engine to idle very well. This is because the 3 BAR MAP sensor has an output voltage that is 1/3 that of the original MAP sensor at the vacuum of an idle. The Spark Advance table and Volumetric Efficiency (VE) table in the Fiero ECM have numerous columns of data in the range of vacuum at idle with the original MAP sensor. However, with the 3 BAR MAP sensor, the tables get compressed so that the first column of data is far above the vacuum at idle. This makes it very difficult to make changes to improve the idle with the 3 BAR MAP sensor. After the piston broke we backed the boost down to a more reasonable level and switched back to the 2 BAR MAP sensor, with the modification described below.

Turbocharging Normally Aspirated Engines on a Budget

The MAP sensor output ranges from about 0.5VDC to 4.9VDC, its output being pulled down by a resistor in the Fiero ECM. We found that when this voltage goes above about 4.7 VDC, the Fiero ECM shuts off the engine and gives an error code. With a 1 BAR MAP sensor and no boost, this only happens if the MAP sensor had failed. With the turbocharger and the 2 BAR MAP sensor, this happens when increasing the boost, and the pressure bumps up against the upper limit of the MAP sensor. A pressure spike in the intake manifold can occur when the throttle is closed/opened suddenly, particularly in a system without a BOV (Blow-Off Valve), and the boost pressure needs to be a few PSI below the upper limit of the MAP sensor to avoid this problem. For example, without a BOV the boost should be limited to about 12 PSI with the 2 BAR MAP sensor. With a BOV the boost pressure can be set very close to the upper limit of the MAP sensor. For example, with a BOV the 2 BAR MAP sensor, can be used up to about 14 PSI of boost.

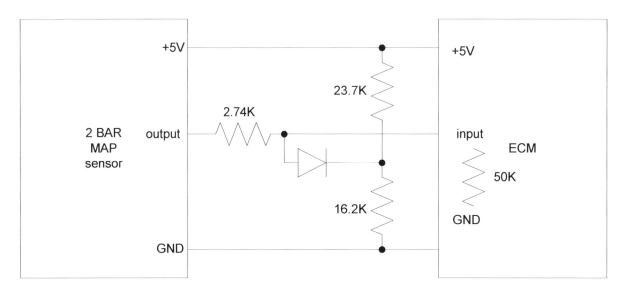

In order to allow the 2 BAR MAP sensor to operate at slightly higher boost, the circuit above was added between the output of the MAP sensor and the input of the Fiero ECM. This circuit allows the 2 BAR MAP sensor to operate normally between 0 and 1 BAR, and then operate on a different slope between 1 and 2 BAR, so that it will actually sense accurately up to about 18 PSI of boost. It also keeps the output voltage from exceeding the limit that causes the Fiero ECM to momentarily shut off the engine and give an error code. This change in slope of the 2 BAR MAP sensor must be coordinated with the Volumetric Efficiency (VE) table in the Fiero ECM for values above 1 BAR. Finally, this change allowed the Fiero to idle well because it responds like it did previously with manifold vacuum (absolute pressure < 1 BAR).

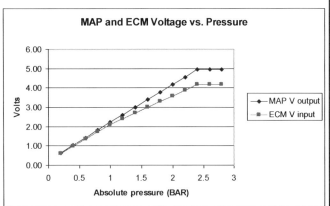

boost (PSI)	absolute (PSI)	absolute (BAR)	MAP V output	ECM V input
-11.76	2.94	0.2	0.64	0.605
-8.82	5.88	0.4	1.03	0.978
-5.88	8.82	0.6	1.42	1.350
-2.94	11.76	0.8	1.82	1.723
0	14.7	1	2.21	2.095
2.94	17.64	1.2	2.60	2.405
5.88	20.58	1.4	3.00	2.699
8.82	23.52	1.6	3.39	2.994
11.76	26.46	1.8	3.78	3.288
14.7	29.4	2	4.17	3.582
17.64	32.34	2.2	4.57	3.877
20.58	35.28	2.4	4.96	4.171
23.52	38.22	2.6	4.96	4.171
26.46	41.16	2.8	4.96	4.171

© 2012 Robert G. Wagoner
All rights reserved.

Turbocharging Normally Aspirated Engines on a Budget

The First Problem: Buckled Exhaust

The first minor failure was that a section of the "Y" exhaust pipe collapsed and buckled, as shown in the photograph. Not surprisingly, this caused a noticeable loss of power. This failure was caused by a combination of fiberglass exhaust wrap around this pipe allowing the pipe to get too hot from the heat from the exhaust, coupled with some mechanical pressure from the turbocharger bumping against the shock tower when the engine moved from the torque. We had originally used

the fiberglass exhaust wrap to shield nearby wiring from the heat of the exhaust pipes. After this failure we repaired the exhaust and added a heat shield instead of the fiberglass exhaust wrap. We also moved the turbocharger a bit further away from the shock tower to eliminate the interference. Not long afterward we added the intercooler, at which time we replaced all motor mounts and added an extra motor mount to keep the engine from moving. These changes solved this problem.

Spark Knock

Refer to the section in Chapter 2, page 18, for background information on spark knock. Here is a list of the improvements we made on the 3.4L engine to counteract most of these factors that cause spark knock.

- Added a head spacer that reduced the compression ratio.
- Cleaned all carbon deposits when we removed the heads and valves during engine installation.
- Installed spark plugs that are one range colder than stock
- Retarded the ignition timing
- Used lower temperature thermostat to keep engine cooler
- Added an intercooler to reduce the inlet air temperature after the turbocharger
- Decreased Air-to-Fuel ratio down to 13:1 under acceleration, making the mixture very rich.
- Used premium fuel, which has a higher self-ignition temperature.

Colder Plugs

The temperature of the spark plug tip is important. It must be hot enough to keep itself clean, but it must be kept below the ignition temperature of the air/fuel mixture. If the spark plug tip gets too hot, it will cause "pre-ignition" to occur. In this particular case, ignition timing cannot correct this problem, since the fuel is being ignited by the hot tip of the spark plug instead of by the spark itself. Refer to the section in Chapter 2, page 18, for background information on colder plugs. For reference ACDelco R42CTS or R42TS plugs are the standard plugs for stock V6 Fieros. These cross to NGK UR5 plugs, which NGK recommends for all 2.8L V6 Fieros and also for all 3.4L V6 Camaros.

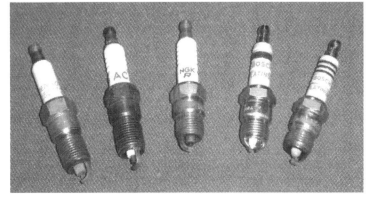

Turbocharging Normally Aspirated Engines on a Budget

Reducing Compression Ratio

The original compression ratio of the 2.8L Fiero engine is 8.9:1. The original compression ratio of the 3.4L Camaro engine is 9.0:1. For the purpose of turbocharging an engine, it helps to reduce the compression ratio, which effectively allows cramming more air/fuel into the cylinder before having trouble with detonation. When we decided to switch to the bigger turbocharger, we wanted to reduce the compression ratio to allow more boost pressure without detonation. We found a few companies that make steel spacers to go under the heads or copper head gaskets in various thicknesses to reduce the compression ratio. We decided to use a steel spacer, 0.075" thick, directly against the block, and an ordinary head gasket. Sources for such custom components are few and far between. We got these from Innovative Machine & Supply in St. Louis, MO. These head spacers are made from 0.075" cold rolled steel. These are installed using a spray gasket sealer, Hylomar, between the spacer and the block. Adding this steel spacer reduces the compression from the original 9.0:1 to 7.5:1, as calculated with the formulas shown on page 18.

Fuel injectors

As more air is added to the engine, it needs more fuel to match. This could be done by increasing the fuel pressure, or increasing the fuel injector size, or a combination of both. We increased the fuel pressure from the stock 44 PSI to 46 PSI by adding a washer under the spring in the fuel pressure regulator. Due to the limitation of the fuel pump pressure vs. flow curve, it is not possible to increase the fuel pressure enough to get the necessary flow through the original fuel injectors. Therefore it is also necessary to change the fuel injectors, and in fact the change to the fuel pressure is almost insignificant in comparison.

> **Fuel Vaporization**
>
> The Fiero V6 utilizes a MFI (Multi-port Fuel Injection) scheme. All fuel injectors in each bank are electrically connected in parallel, so they all open at the same time, once every revolution of the engine. Since it takes two revolutions between each firing, the fuel injectors have sprayed fuel two times between each firing. While the fuel sits in the intake manifold it has time to vaporize before it is actually drawn into the cylinder on the intake stroke. The fuel injector sprays fuel onto the backside of the closed intake valve, which is hot, improving vaporization.

Before calculating fuel injector size, it is important to understand the fuel pressure regulation system on the Fiero. It consists of a fuel pressure regulator that is referenced to the manifold pressure. With vacuum of the engine the fuel pressure drops accordingly. When we add a turbocharger to the Fiero, this fuel pressure will increase to match the boost pressure. This causes the fuel pressure to remain constant with respect to the manifold pressure, not based on outside ambient pressure. This makes a fixed amount of pressure drop between the fuel rail pressure and the intake manifold pressure. The reason this is important is the fuel pressure effects fuel injector fuel flow rate. This fuel pressure regulator design compensates for this, simplifying fuel injector selection and making tuning easier.

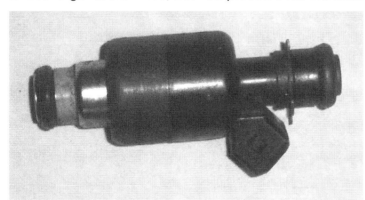

© 2012 Robert G. Wagoner
All rights reserved.

Turbocharging Normally Aspirated Engines on a Budget

The fuel injectors have a rated flow rate that is usually measured in LB/hr. This flow rate of the fuel injector is based on 43.5 PSI of pressure drop across the fuel injector, with the fuel injector duty cycle being 100%. Our fuel pressure regulator is set at 46 PSI, so when a fuel injector is used in the Fiero, the actual fuel flow will be slightly higher than its rated value, by a factor of $(46/43.5)^{0.5}$ = 1.03. Since this factor is small, to avoid confusion, this factor will be omitted in the calculations below. This will give a little extra margin in the fuel injector capability beyond the calculated values. If one wanted to include this factor in the calculations, simply multiply the calculated HP below by a factor of 1.03.

The formula for calculating fuel injector size is shown on page 22. We purchased three different sets of fuel injectors for this turbocharging project: 19 LB/hr, 24 LB/hr, and 30 LB/hr. Originally we used the 19 LB/hr fuel injectors with the small turbocharger up to 7 PSI of boost. We measured a maximum of 187 HP in this configuration. This is really the upper limit of these fuel injectors, as calculated below.

Fuel injector duty cycle = 187 x 0.6 / 6 / 19 = 0.98 = 98.4 % duty cycle

We switched to the 24LB fuel injectors when we decided to add the intercooler and push the boost up above 7 PSI. Turning that formula around, below is a calculation of the expected maximum HP for the 24 LB/hr fuel injectors at 100% duty cycle. It is usually recommended to allow some margin to keep fuel injector duty cycle below 100%. In the calculation below 0.6 is used for BSFC parameter, which is the factor that provides margin.

Maximum HP = 24 x 6 x 100% / 0.6 = 240 HP

We used the 24 LB fuel injectors with the small turbocharger, and measured a maximum of 235 HP. At this level, the fuel injector duty cycle was approaching its limit, as calculated below.

Fuel injector duty cycle = 235 x 0.6 / 6 / 24 = 0.98 = 97.9 % duty cycle

When we switched to the large turbocharger, we also switched to 30 LB/hr fuel injectors, scaling up our power expectations from 235 up to 300, which is very close to the ratio between 24 LB/hr and 30 LB/hr. After many tests, iterations, and failures, in the end this setup was adjusted back down to 11 PSI boost pressure to ensure it would be very reliable, and back to the 24 LB/hr fuel injectors.

Fuel Pump

The table at the right shows the performance of the stock fuel pump at 12VDC. The typical voltage of an automobile is closer to 14VDC while it is running, and the pump spins faster at higher voltage, creating more pressure and flow than at 12VDC. For this reason it is best to verify the pump has plenty of flow at the desired pressure at 12VDC, to ensure it has even more margin at the typically higher operating voltages. At 50 PSI the fuel pump flow is 36 gal/hr. Gasoline weighs roughly 6.2 LB/gal, more or less depending upon the additives and alcohol content. To be conservative use BSFC = 0.6, resulting in this fuel pump being capable of 372 HP. This is far in excess of the amount necessary here, and therefore the stock fuel pump will be used.

Pressure (psi)	fuel pump (gal/hr)	fuel pump (amps)
0	56	2.6
10	52	3.2
20	48	4.0
30	44	4.8
40	40	5.7
50	36	6.6
60	32	7.6
70	28	8.6
80	22	9.7
90	12	10.8
100	0	12.5

Turbocharging Normally Aspirated Engines on a Budget

Adding an Intercooler

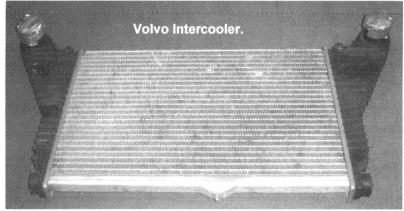
Volvo Intercooler.

For tests up to 7 PSI of boost with the small turbocharger, the turbocharger fed directly into the engine without an intercooler. As we decided to increase the boost above that level, we knew we would need to add an intercooler. The only location with enough room for a large intercooler is the trunk. We found that an intercooler from a Volvo Turbo fits nicely in the trunk, leaving room for a fan to be mounted behind it. We used an extra radiator fan from a Fiero to suck air through the intercooler and blow it out the back of the trunk. We removed the carpet from the trunk and cut a bunch of large holes in the front, sides, and back walls of the trunk for airflow and plumbing. A few pipes later, intercooler complete. We measured this setup to have 1 PSI pressure drop at 330 CFM with an intercooler efficiency of 71%. At this point we started increasing the boost again.

More Torque Requires Stronger Support

When we decided to increase the boost beyond 7 PSI, particularly with our goal being 20 PSI, we believed it would be important to beef up a few items that were still stock.

Clutch: We upgraded to a Centerforce Dual Friction Clutch. We did not actually have any trouble with the stock clutch, because we never gave it a chance. According to Centerforce, these generate up to a 90% increase in holding capacity over stock systems. We didn't have any clutch slippage problems with this clutch, clear up to the point we had pushed the boost high enough to break the 3.4L engine. The Centerforce Dual Friction clutch may not have been necessary at the recommended 11 PSI. The Centerforce Dual Friction clutch was added as we decided to find the limit of this 3.4L engine, and increased the boost level far beyond 11 PSI.

"Ultimate Dogbone"

Engine Mounts: The engine and transaxle mounts were old, but still solid, so they were not changed previously. As we had increased the torque by adding 7 PSI of boost, we started to notice more engine movement during acceleration. At this point in time we decided to add an intercooler, which we mounted in the trunk. This made engine movement more of a concern, because of the plumbing between the engine and the intercooler. We knew that when we went higher in boost, the engine and transaxle mounts needed to be as strong as possible, so we replaced them all. We also replaced the engine strut support motor mount on the passenger's side of the engine with the "Ultimate Dogbone" which can be purchased online at Fiero Store or West Coast Fiero. By the time the boost level was up to 16 PSI with the large turbocharger, we found we still had too much engine movement, occasionally causing one of the intercooler piping connections to pull apart. So we added an additional brace directly between the top of the engine and the trunk on the driver's side, somewhat similar to the engine strut support motor mount on the passenger's side. This reduced the engine movement enough that we no longer had any trouble with the plumbing between the engine and the intercooler.

© 2012 Robert G. Wagoner
All rights reserved.

Turbocharging Normally Aspirated Engines on a Budget

Addding a BOV

When we decided to go above 7 PSI and add an intercooler, we also added a Blow-Off Valve (BOV) from a Volvo Turbo to the output of the turbocharger, before the intercooler. The BOV is controlled by the manifold vacuum, so that when the driver lets off the throttle suddenly, the BOV opens to release the excess pressure from building up at the outlet of the turbocharger. Since this opens the intake to another source of air, an extra air filter was added to the outlet of the BOV, as shown in the photograph at the right.

BOV

New Gauges for Tuning

For tuning I used an Air-to-Fuel Ratio Gauge and a Boost Pressure Gauge. These are both very valuable for monitoring the performance of the turbocharger. These gauges could be mounted in the car permanently with an "A-Pillar Dual Gauge Pod" from the Fiero Store, which looks quite professional, or they could be removed after tuning is complete. After tuning is complete they are less useful, so I do not mount them permanently. In my case, I move them from car to car as I am tuning the next one. A third gauge would also be valuable for tuning, an exhaust gas temperature gauge. This was the first car I had turbocharged, and at this time I did not have an exhaust gas temperature gauge.

Reprogramming the Fiero ECM

Tuning the system took more time than installing the turbocharger. At first the "check engine" light kept coming on, the Block Learn Multiplier (BLM) parameters were all out of whack, the Air-to-Fuel Ratio (AFR) was much too lean at most operating points, there was spark knock, etc.

The procedure to adjust this is to:
1) Drive the car and collect data. This takes about 10 minutes.
2) Go home, analyze the data, and reprogram the ECM. This takes about 2 hours.
3) Go back to step 1 about 40 times. This takes months!

To monitor the status of the Fiero while it is running, it takes a special interface to the engine diagnostic connector located behind the panel on the console between the seats. There are many engine scanners that can be purchased to perform this function, such as the "Auto X Ray Pro Pack", which sells for $399. However, I like to do things myself, not only to save money, but also because it gives me a deeper understanding of the way things work. So instead of purchasing one of these engine scanners, I built my own version. It was really not as hard as it sounds, because of a program called WinALDL. WinALDL is free software that makes it easy to use a laptop computer to communicate with the Fiero ECM

Fiero ECM with cover removed to reveal EPROM

through the ALDL connector. This requires a special ALDL interface cable to connect between laptop computer and the ALDL connector. This special interface cable only requires one transistor and three resistors so it is extremely easy to build. The instructions are described clearly on the website where WinALDL is found.

© 2012 Robert G. Wagoner
All rights reserved.

Turbocharging Normally Aspirated Engines on a Budget

As the car is being driven, the Fiero ECM will adjust the BLM to achieve the proper AFR when it goes into closed loop mode. So we only need to get close with the VE table, and the Fiero ECM will do the rest. By using a laptop computer to monitor the BLM values, we get new valuable information with each test run. Then we go back and reprogram the ECM to get closer to the proper values, so that the Fiero ECM does not have to make as much adjustment to achieve the proper Air-to-Fuel Ratio.

To reprogram the EPROM in the Fiero ECM requires an UV eraser and an EPROM programmer. These are available on eBay for less than $30, in case you don't already own one. ☺

A company called "CATS" provides software for tuning the Fiero which works great. This software reads the EPROM and prints out tables of the parameters, making it easy to study and change the parameters.

It is convenient to have more than one EPROM. The EPROMs are Intel part number D2732A. I would usually program three EPROM's at a time, each with a slightly different fuel map and spark map, and test all three while we were out driving the car to find out which one worked the best. Then the best one would be the baseline for the next three versions. With many, many iterations, we slowly worked to a version that worked well. After much effort, the engine idled and ran great. In this chapter I am going to jump directly to the final parameter values we found that worked well. In the next chapter I will go into a little more detail about the process of tuning.

Besides the normal tuning, the original Fiero rev limiter was always a problem when Autocross racing, because after it turned the fuel off at 5851 RPM, it kept the fuel off until the engine speed dropped to 4008 RPM. That behavior makes for a very jerky ride. The rev limiter action was improved with the following changes.

> Fuel cutoff changed from 5851 to 6012 RPM
> Fuel resume changed from 4008 to 5904 RPM.

The tables and graphs on the following pages include the final results of Volumetric Efficiency (VE) tables and Spark Advance tables. The calculation of fuel injector pulse width is based on the VE table, along with a number of other variables, including a constant that gets multiplied to the total. This constant, sometimes referred to as BPC (Base Pulse Constant), is listed below the table as a "VE multiplier" included along with each VE table. This VE multiplier directly affects the amount of fuel that goes into the engine. One common use for this variable is to adjust for changes in the fuel injector size. For example, if the fuel injector size was increased by 10%, this variable could be decreased by 10% to compensate.

The final tuning values shown on the following pages worked well on our 3.4L engine with 11 PSI boost pressure and the 24 LB/hr fuel injectors. After turbocharging five Fieros so far, I can testify that every installation performs a bit differently, and the values in the tables below would be a good place to start, but probably not perfectly tuned for any other Fiero.

Turbocharging Normally Aspirated Engines on a Budget

MAP voltage	0.511	1.040	1.569	2.099	2.628	3.157	3.686	4.215	4.745
BAR	0.4	0.6	0.8	1.0	1.2	1.4	1.6	1.8	2.0
800 RPM	6.3	18	20.7	27.3	32.4	37.1	42.2	47.3	52.3
1200 RPM	9.4	21.1	23.4	29.7	34.8	39.8	44.9	49.6	54.7
1600 RPM	14.5	24.2	26.6	32	37.1	42.2	47.3	52.3	57
2000 RPM	19.5	27.3	30.1	34.8	39.8	44.5	49.6	54.7	59.8
2400 RPM	21.9	30.5	33.6	37.1	42.2	47.3	52	57	62.1
2800 RPM	23.8	33.2	35.9	39.5	44.5	49.6	54.7	59.4	64.5
3200 RPM	27	35.2	38.3	42.2	46.9	52	57	62.1	67.2
3600 RPM	29.7	36.3	39.8	44.5	49.6	54.3	59.4	64.5	69.5
4000 RPM	33.2	38.7	41.4	46.9	52	57	61.7	66.8	71.9

VE table with 2 BAR MAP sensor and 24 LB fuel injectors, VE multiplier = 208

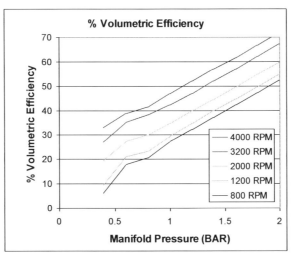

A plot with five lines of data from VE table for small turbo, 2 BAR MAP sensor, and 24 LB fuel injectors

MAP voltage	0.511	0.776	1.040	1.305	1.569	1.834	2.099	2.363	2.628	2.892	3.157	3.422	3.686	3.951	4.215	4.480	4.745
BAR	0.4	0.5	0.6	0.7	0.8	0.9	1	1.1	1.2	1.3	1.4	1.5	1.6	1.7	1.8	1.9	2
600 RPM	28.1	28.1	28.1	28.1	28.1	13.4	5.6	2.5	0	0	0	0	0	0	0	0	0
800 RPM	32	32	32	32	28.8	20	13	9.8	7	3.9	1.1	0	0	0	0	0	0
1000 RPM	35.9	35.9	35.9	35.9	28.8	20	13	9.8	7	3.9	1.1	0	0	0	0	0	0
1200 RPM	40.1	40.1	40.1	35.9	33	26	16.9	14.1	10.9	7.7	4.9	1.8	0	0	0	0	0
1400 RPM	41.8	41.8	40.1	35.2	33	27.1	17.9	14.8	12	8.8	6	2.8	0	0	0	0	0
1600 RPM	46.1	46.1	40.1	35.9	34.1	27.1	19	16.2	13	9.8	7	3.9	1.1	0	0	0	0
2000 RPM	46.1	46.1	41.8	35.9	34.1	28.8	23.9	20.7	17.9	14.8	12	8.8	6	2.8	0	0	0
2400 RPM	46.1	46.1	46.1	40.1	35.2	29.9	26	22.9	20	16.9	14.1	10.9	8.1	4.9	2.1	0	0
2800 RPM	48.2	48.2	46.1	40.1	36.9	29.9	27.1	24.3	21.1	17.9	15.1	12	9.1	6	3.2	0	0
3200 RPM	45	45	41.8	40.1	36.9	33	28.8	25.7	22.9	19.7	16.9	13.7	10.9	7.7	4.9	1.8	0
3600 RPM	41.8	41.8	40.1	39	35.2	28.1	28.1	25.0	22.1	19	16.2	13	10.2	7	4.2	1.1	0
4000 RPM	45	45	45	39	35.2	29.9	28.1	25.0	22.1	19	16.2	13	10.2	7	4.2	1.1	0
4400 RPM	48.2	48.2	46.1	39	35.9	33	29.9	27.1	23.9	20.7	17.9	14.8	12	8.8	6	2.8	0
4800 RPM	49.9	49.9	46.1	39	35.9	33	29.9	27.1	23.9	20.7	17.9	14.8	12	8.8	6	2.8	0

Spark Advance table for small turbocharger and 2 BAR MAP sensor

© 2012 Robert G. Wagoner
All rights reserved.

Turbocharging Normally Aspirated Engines on a Budget

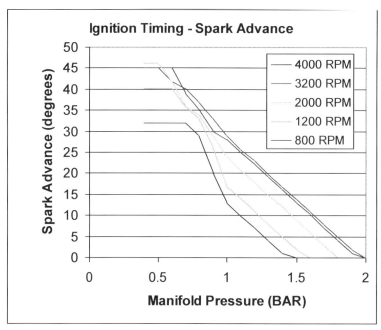

A plot with five lines of data from Spark Advance table for small turbocharger and 2 BAR MAP sensor

© 2012 Robert G. Wagoner
All rights reserved.

Turbocharging Normally Aspirated Engines on a Budget

The Recommended Configuration

From this experience, we learned in order to make a reliable setup when adding a turbocharger to a Fiero with a GM V6 3.4 L 60° engine, use the Garrett T04-"50" turbocharger with an intercooler, set the boost at 11 PSI, use 24 LB/hr fuel injectors, and a 2 BAR MAP sensor. Going beyond that to higher boost pressure is difficult for many reasons, and stresses the engine beyond its reliable limit.

Parts List and Costs

Here is a list of the parts and costs for the recommended installation, not including experimental parts, special tools, or things we broke and replaced, such as the piston. Many of the parts were purchased on eBay, either new or used. These are sorted by cost.

"Recommended" parts		
source	description	cost
ebay	T3/T04 turbocharger, new no-name	$ 235.00
ebay	2" intercooler piping with silicone rubber connectors & clamps, ne	$ 106.98
Innovative Machine & Supply	2 of 0.075" GM173 Head spacer and 1 Hylomar Aerosol	$ 101.30
** local muffler shop	Exhaust down pipe with flange for turbocharger	$ 100.00
ebay	HKS blow off valve, new	$ 64.99
local auto parts store	2 BAR MAP sensor	$ 64.57
ebay	8 of Fuel Injectors 24 lb LT1 TPI Corvette, used	$ 60.00
West Coast Fiero	WCF Ultimate Dogbone engine mount	$ 55.00
ebay	Volvo intercooler, used	$ 42.02
local auto parts store	2 head gaskets for use with 0.075" steel head spacers	$ 21.51
ebay	oil feed kit	$ 18.90
local auto parts store	6 of NGK UR6 spark plugs	$ 10.53
local auto parts store	small red air filter for PCV vent	$ 10.44
ebay	air filter, new	$ 9.98
local auto parts store	oil return line (2 feet of 1/2" fuel compatible hose) & 2 hose clamp	$ 9.01
local hardware store	oil return fitting (1/2" brass female hose barb to pipe thread)	$ 7.86
local hardware store	Teflon tape & hose clamps	$ 6.96
ebay	small air filter for BOV vent to ambient	$ 3.88
local hardware store	1/4" Brass T and 1/4" brass nipple	$ 3.87
local auto parts store	3' hose 5/32" for coolant feed	$ 2.13
	grand total	$ 934.93

** This is an estimate for cost of custom exhaust work at a local shop. I make my own exhaust down pipes.

Innovative Machine & Supply	(800)-736-1875 {St Louis, MO}	
West Coast Fiero	(310)-305-4111 {Marina Del Rey, CA}	

Breakage Cost

During this project, the extra cost of components used for experiments and components broken was an extra $890. The list of major components includes a second turbocharger and a damaged piston, which resulted in a complete engine rebuild, caused by pushing boost to above 20 PSI.

Turbocharging Normally Aspirated Engines on a Budget

> **_Warning_**
> _Modifications in this book may not be legal for street cars. Check local and state laws before making any changes to your car._
>
> _Modifying your engine will void any factory warranties._
>
> _The author does not recommend that anyone make any modifications to their car. The author is not responsible for any loss, damage, or injury caused by any modifications anyone makes to their car._

Expect Breakage
- *I have had multiple major failures when modifying engines to increase power.*
- *I have learned to start a project like this with an expectation of major damage, and I feel lucky if that doesn't happen.*
- *I have done the most damage when I decided at the end to bump up the boost more and more until I "find the limit".*

Reliability
- *Minimize the number of modifications – every change reduces reliability.*
- *Start with a good-running, low-mileage engine.*
- *Ensure every modification is done carefully and correctly.*
- *Plan on a conservative power increase – don't push it.*

> *This book should be used in conjunction with a factory service manual, or a repair manual for your specific vehicle from GM, Chilton, or Haynes, which will include safety procedures, engine rebuilding information, torque specifications, cleaning, etc.*

4 Turbocharging a 1.8L 1ZZ-FED Engine in a 2001 MR2 Spyder

This is another example of turbocharging a normally aspirated engine for less than $1000. Compared to the first example, this second example utilizes a much different fuel control, a Piggyback Fuel / Ignition Controller (FIC). The previous chapter focused more on the hardware, and much of that detail will not be repeated in this chapter. This chapter will provide more detail about turbocharger system design and tuning and less on the hardware installation. This chapter combined with the last one should provide a fairly solid basis for turbocharging many other cars. The subsequent chapters will tend to get shorter as they provide less and less new information.

To avoid duplication of ideas, paragraphs, photographs, and pages, I will occasionally be referring to previous sections and chapters in this book. If a person reads the entire book, this will make it more interesting for them, not having to read the same stuff over and over. It would be difficult to get the overall picture if a person only reads this one chapter about a single car. With this in mind, even if you are just trying to turbocharge this particular engine, I suggest you first read chapters 1 through 4 to get some background information before starting this chapter.

For variety, I have written this chapter directly from the "notebook" I made as I did this work. My "notebook" for this project is a Word document 131 pages long. This is organized in the sequence I did the work, first with designs and calculations, followed by some test data of the car before it was modified, then documentation of the changes made to turbocharge it, and finally the tuning of the new turbocharged system. I cut the length down from 131 pages to 23 pages of the most important information for this chapter of this book.

© 2012 Robert G. Wagoner
All rights reserved.

Turbocharging Normally Aspirated Engines on a Budget

Part 1: Turbocharger System Design

The goal of this section is to explain how to consider all options and choose the best option. One important result of this work will be a system block diagram like below.

Turbocharger System Block Diagram

To turbocharge this 1.8L engine, the inlet air is pulled from the rear of the engine compartment in an area isolated from the engine heat by heat shields. This is where the new air filter is located. From the air filter the air will feed next into the turbocharger compressor inlet, and then to the intercooler. The intercooler is also located in the rear of the engine compartment in an area isolated from the engine heat by heat shields. From the intercooler it flows to the Blow-Off Valve (BOV), which is installed onto the side of the intake piping in a way to minimize disturbance of the airflow. Next the air flows to the MAF sensor, and finally into the throttle body, which is attached to the intake manifold, as shown below. The BOV has its own small air filter attached directly to it, because when it is open it creates a source of air to the system that needs to be filtered. Since the MAF sensor is located after the BOV, the air "leak" caused by the BOV does not have an adverse influence on the ability of the MAF sensor to control the Air-to-Fuel Ratio (AFR). This allows the BOV to vent through its own air filter directly to the outside air.

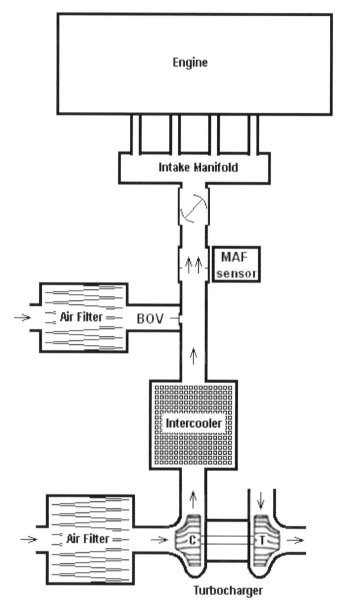

Mass Air Flow (MAF) Sensor Location

The engine in the previous chapter did not have a MAF sensor, so this is the first chance I have to address this particular sensor in detail. The MAF sensor measures the airflow and sends that information to the ECM, which is the primary controlling input to an algorithm that determines the fuel injector pulse width to create the appropriate Air-to-Fuel Ratio. The MAF sensor should be located past the intercooler close to the throttle body to avoid any lag or errors in its sensing that can be caused by delays and pulsations through the intercooler and piping.

Also it is most important that a Blow-Off-Valve (BOV) is located in a location that it does not cause a leak that bypasses the MAF sensor when it is open. The BOV may be located before or after the intercooler, as long as it is located between the turbocharger outlet and the MAF sensor. When located after the intercooler, as shown in the diagram on the previous page, there should be a long, straight pipe between the BOV and the MAF sensor, at least 12" long, to help straighten and smooth the air as it flows into the MAF sensor.

In contrast, in cases where the MAF sensor is located before the turbocharger compressor inlet, the BOV must be vented back into the intake plumbing between the MAF sensor and the turbocharger compressor inlet. Chapter 7 is an example like this, including more details about it.

Cooling and Lubrication

My initial thoughts on turbocharging this engine were about cooling and lubrication.

Turbocharger water feed: the turbocharger can get water from the coolant feed going through throttle body. The coolant feed going through throttle body has minimal value in warm climates, and since I live in Virginia, I plan to remove it from the throttle body and route it to the turbocharger instead. Details of the fittings on the turbocharger itself will be discussed later.

Turbocharger oil feed: the turbocharger can get oil exactly like the previous chapter, with a feed from the oil pressure sensor, and a return directly into the oil pan. This will require a special metric to standard T-fitting for oil sensor. The source and part number can be found in the parts list.

Options Considered

Since the existing fuel control is based on a MAF sensor, this provides a number of options to consider. I considered eight options on various ways to turbocharge this engine.

Fuel System Control Upgrade Options 1 – 8 {copied directly from my notebook}
1) No change to fuel system and fuel injectors. Keep boost low enough that it works with no ECM & MAF changes. Add piggyback FIC for monitoring and adjustments only if necessary.
2) Everything same as 1) except add an electronic boost controller to allow more boost at lower RPM to get more torque.
3) Everything same as 1) except modify existing fuel pressure regulator to increase fuel pressure by ~10% like I did in the Fiero. This might work if the existing fuel injectors are close but just short of handling desired HP. Might need to scale MAF with a slightly larger tube diameter to compensate for the change in fuel flow. .
4) Add a 10:1 rising rate boost fuel pressure regulator FMU and an extra fuel pump , like the 1999 Dodge Neon setup with no piggyback FIC. This would allow higher HP without changing fuel

Turbocharging Normally Aspirated Engines on a Budget

injectors. Would need to add a circuit to limit MAF voltage to 4.75VDC max, so it doesn't cut out or give error code at high airflow.
5) Larger fuel injectors, existing MAF, existing fuel pump. Add piggyback FIC. Would need to adjust MAF scaling to match new fuel injectors, and limit MAF voltage like 4) above.
6) Larger fuel injectors, reuse the existing MAF sensor in a larger tube, calculated size of larger tube to match larger fuel injectors.
7) Boost rising rate fuel pressure regulator FMU & extra fuel pump coupled with piggyback FIC.
8) Add 5th fuel injector in intake manifold, operated only with boost. Use a piggyback FIC to adjust MAF scaling, ignition timing, and to control the operation of the additional fuel injector.

Boost pressure control for Options 1 – 8 {copied directly from my notebook}
1) Nothing more than a wastegate. Simply pick a wastegate with fixed boost pressure control at a low enough level to never hit the limit of existing fuel injectors and MAF, as well as a boost pressure that runs without knocks. Initial guess without calculations: maybe 6-8 PSI without intercooler, or maybe 10-11 PSI with an intercooler.
2) The boost control could be a simple electronic circuit, which would be relatively easy to design. Would need to monitor the MAF sensor, and cut boost when MAF hits 4.75 VDC. This would require setting the wastegate to perhaps 4 - 6 PSI, adding a boost pressure regulator to set desired level to perhaps 8 – 10 PSI, and an electric solenoid to switch this in and out when the MAF gets near its limit. Alternately I could buy an aftermarket electronic boost pressure controller.
3) No electronic boost control required. Same as 1).
4) No electronic boost control required. Same as 1).
5) No electronic boost control required. Same as 1).
6) No electronic boost control required. Same as 1).
7) No electronic boost control required. Same as 1).
8) No electronic boost control required. Same as 1).

ECM changes and/or Piggyback FIC for Options 1 – 8 {copied directly from my notebook}
1) Might not need a piggyback FIC to adjust AFR, since this system is based on a MAF sensor with some margin. Might need a piggyback FIC to adjust timing. It would be handy to buy a piggyback FIC to make measurements and ensure fuel and timing can be adjusted as necessary to make it will work, and if the piggyback FIC is not really required, remove it later.
2) Same as 1).
3) Same as 1).
4) Same as 1).
5) Rescale the fuel injector pulse widths by the ratio of old fuel injector size / new size. Add a MAF voltage limiter to 4.75VDC max, so it doesn't cut out or give error code at high airflow. Use pressure to modify MAF signal, starting at 15 PSI, dropping 7% per PSI.
6) Same as 1).
7) Add a MAF voltage limiter to 4.75VDC max, so it doesn't cut out or give error code at high airflow. Modify MAF scaling as soon as boost starts, dropping 7% per PSI boost.
8) Similar to 5)

Benefits/drawbacks of Options 1 – 8 {copied directly from my notebook}
1) Lowest cost. This will be the lowest HP option, due to fuel limitation, with no upgrade to fuel flow capacity. This has fewest changes, hence less likely to have loose connections, and probably the most reliable solution. Piggyback FIC bypass option improves availability.
2) Similar to 1) except slightly more torque with slightly more cost. Drawback is hand designed and hand built electronics, which have risk of reliability problems.
3) Similar to 2) except slightly more HP with slightly more cost. I expect the 10% higher fuel pressure to be compensated in closed loop by ECM/oxygen sensor.

Turbocharging Normally Aspirated Engines on a Budget

4) Allows higher HP without changing fuel injectors. Always runs rich with boost (probably a big drawback, because it might mess up the closed loop Air-to-Fuel Ratio table). Hence option 7) preferred. Benefit adding extra fuel in-line fuel pump adds capacity to existing fuel pump. Drawbacks: High fuel pressure is not desirable due to the safety risk. Also the FMU's are not reliable.
5) Adding FIC adds risk of intermittent wiring problems. Another risk is the capacity of the existing fuel pump is unknown, so could go lean. Still has MAF limit problem. Even better if the piggyback allows adjustment of ignition timing.
6) Benefit is no FIC required, hence no risk of intermittent wiring problems. An important benefit is this allows higher maximum HP than stock fuel system. Risk is getting the size of the tube for MAF sensor just right. Another risk is the capacity of the existing fuel pump is unknown, so could go lean – need to check this first. If it runs out of fuel, might need to add an extra fuel pump and external regulator. Would test initially without a piggyback FIC, but if there was some problem, then might need to add a FIC to adjust ignition timing & MAF scaling.
7) This keeps the ECM tuning exactly the same anytime there is no boost. Benefit adding extra fuel in-line fuel pump adds capacity to existing fuel pump. The piggyback FIC should make it possible to correct AFR. Drawbacks: High fuel pressure is not desirable due to the safety risk. Also the FMU's are not reliable. Adding piggyback FIC adds risk of intermittent wiring problems.
8) Lower reliability due to extra fuel injector and FIC is required to control it. Possible air/fuel mixture problems, lean cylinders, fuel vaporization problems, fuel puddle in intake manifold, etc. Might be fun to experiment with on an engine I don't mind breaking, but sounds too risky for this project.

My first choice was unsuccessful, and wasted about a month of time. To make a long story short, first I tried option 6, because it seemed simple and had potential to be a very low cost upgrade to ensure there was plenty of potential for additional boost. Supra 7MGE fuel injectors flow 295cc/min (yellow) or 315cc/min (green) at 36 PSI, and are high impedance. They looked like they might fit from photographs, so I purchased a set of used Supra 315cc/min (green) fuel injectors for $55. Other fuel injectors that are physically the right size were too expensive. While I waited for the new fuel injectors to arrive, I built a larger MAF tube, so I could install them together at the same time. I will skip all the detailed calculations of fuel flow and MAF tube scaling and jump directly to the problem. When I installed the Supra fuel injectors, they did not fit properly because they are larger diameter than the 1ZZ fuel injectors, and they end up wedging against the intake manifold in a way to cause a fuel leak. This process took about a month to rule out option 6.

My second choice, option 3, was also unsuccessful, and delayed the project by another week, as the work took an entire weekend. Again, because it seemed simple, I wanted to figure out if I could modify the fuel pressure regulator like I did on the Fiero to increase the fuel pressure to get more fuel flow (see chapter 3). After removing the fuel tank from the MR2 Spyder and disassembling the fuel pump/pressure regulator assembly, I decided this fuel pressure regulator could not be modified this way and survive with certain reliability. So I put it all back together with no modifications.

After these two unsuccessful attempts at modifications, I decided to do some extra tests on the existing fuel system to see how much performance I could get from option 1 (simply use the existing fuel injectors, existing fuel pump, and existing fuel pressure).

Installing AEM FIC: Test Data to Decide on the Best Option

Even in cases where no piggyback FIC is required, it would be nice to get an FIC just for monitoring. I decided to purchase a piggyback FIC. The lowest cost piggyback FIC that has the necessary features is the AEM Fuel / Ignition Controller part number 30-1910, which cost $314 on eBay in 2010. The piggyback FIC will always be beneficial in adjusting timing in the presence of boost if that is necessary. This also provides other possible uses, above and beyond those

Turbocharging Normally Aspirated Engines on a Budget

functions they are traditionally used for. For example, the FIC can fix the downstream oxygen sensor, which I would like to eliminate if I remove the catalytic converter. My original plan was to remove the piggyback FIC if no adjustments were necessary. However, in the end the piggyback FIC was necessary, so it stayed in the system. The table below shows the wiring used to connect this AEM FIC into the 2002 MR2.

AEM FIC 30-1910	2001 MR2 ECM wiring
Pin / Name / Wire Color / Wire Marking /Intercept or Tap	Pin / Color
1 Fuel injector 1 input Dk Blue INJ 1 IN Intercept	D1 B-W
2 Hall style sensor Cam 2 output Yellow CAM2 HALO + Intercept	
3 Hall style sensor Cam 1 output Yellow CAM1 HALO + Intercept	
4 Power GND Black PWR GND Tap	D31 W-B
5 Signal GND Black SIG GND Tap	C17 BR
6 TPS input White TPS + Tap	C23 Y-G
7 Hall style sensor Crank input Green CRK HALI + Intercept	
8 Mag style Crank sensor negative input Green CRK MAGI - Intercept	C24 B through 33K resistor
9 Mag style sensor Cam 2 positive output Yellow CAM2 MAGO + Intercept	
10 Mag style sensor Crank negative output Green CRK MAGO - Intercept	
11 Mag style sensor Cam 1 negative output Yellow CAM1 MAGO - Intercept	
12 Fuel injector 2 input Dk Blue INJ 2 IN Intercept	D2 B
13 User switch input White LOG IN NA	
14 Hall style Crank sensor output Green CRK HALO + Intercept	
15 Power GND Black PWR GND Tap	D31 W-B
16 Ignition power Red IGN PWR Tap	A8
17 Analog B out Orange	
18 Aux Input Brown	
19 Mag style crank sensor positive input Green CRK MAGI + Intercept	C24 B through 33K resistor
20 Mag style Cam 2 sensor negative output Yellow CAM2 MAGO - Intercept	
21 Mag style Crank sensor positive output Green CRK MAGO + Intercept	
22 Mag style Cam 1 sensor positive output Yellow CAM1 MAGO + Intercept	
Pin Name Wire Color Wire Marking Intercept/Tap	
1 Bank 2 oxygen sensor modifier Pink O22 + Tap	C9 W
2 Analog B in Orange	C21 B
3 Analog A out Grey	
4 Hall style Cam 2 sensor input Yellow CAM2 HALI + Intercept	
5 Mag style Cam 2 sensor positive input Yellow CAM2 MAGI + Intercept	
6 Analog A in Grey	D27 B (knock sensor)
7 Mag Style Cam 2 sensor negative input Yellow CAM2 MAGI - Intercept	
8 Hall style Cam 1 sensor input Yellow CAM1 HALI + Intercept	
9 Mag style Cam 1 sensor positive input Yellow CAM1 MAGI + Intercept	
10 Mag style Cam 1 sensor negative input Yellow CAM1 MAGI - Intercept	
11 Fuel injector 5 input Dk Blue INJ 5 IN Intercept	
12 Fuel injector 3 input Dk Blue INJ 3 IN Intercept	D3 L
13 Banks 1 oxygen sensor modifier Pink O21 + Tap	C12 B
14 MAF signal input Dk Blue MAF IN + Intercept	C11 V
15 MAF signal output Black MAF OUT + Intercept	C11 V
16 switched 12V output White	
17 Fuel injector 6 output Dk Blue INJ 6 OUT Intercept	
18 Fuel injector 5 output Dk Blue INJ 5 OUT Intercept	
19 Fuel injector 4 output Dk Blue INJ 4 OUT Intercept	D4 W
20 Fuel injector 3 output Dk Blue INJ 3 OUT Intercept	D3 L
21 Fuel injector 2 output Dk Blue INJ 2 OUT Intercept	D2 B
22 Fuel injector 1 output Dk Blue INJ 1 OUT Intercept	D1 B-W
23 Fuel injector 6 input Dk Blue INJ 6 IN Intercept	
24 Fuel injector 4 input Dk Blue INJ 4 IN Intercept	D4 W

© 2012 Robert G. Wagoner
All rights reserved.

Turbocharging Normally Aspirated Engines on a Budget

Removal of catalytic converter to make room for the turbocharger required faking out the downstream oxygen sensor with an RC circuit to delay the timing of the pulses, in order to avoid code P0420 errors – Oxygen Sensor Catalyst Efficiency Low. Effectively tie the downstream oxygen sensor ECM input directly to upstream oxygen sensor, through a resistor and capacitor with the circuit below.

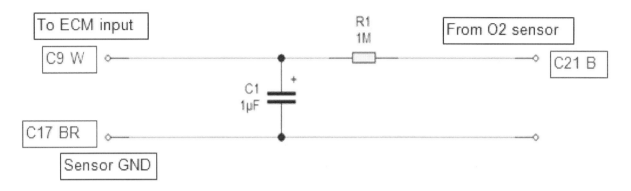

Initial measurements

My plan was to install the AEM FIC to monitor a number of parameters before I installed a turbocharger, and to use this test data to help decide the best option. For example, my first test was to measure fuel injector duty cycle, absolute pressure, and MAF voltage at WOT from 1500 rpm to 6500 rpm in 2nd gear. This test will help show how much margin there is to add boost with existing fuel injectors, existing MAF, and existing ECM. Next I planned to modify the MAF sensor to remove it from the air filter box, and use the AEM FIC to verify it still operated the same as before the modification. This was a necessary step to allow for installing an intercooler in the system. After that I planned to insert the intercooler into the intake plumbing and retest, to determine pressure drop in intercooler, all before adding the turbocharger. Also I wanted to see the oxygen sensor waveforms with an oscilloscope, just to ensure the circuit shown in the schematic diagram above was working properly. All this was done, and test results are shown below and on the following page.

- DC voltage for trip of MAF sensor is 4.9VDC, which scales to 271 gm/sec.
- Front oxygen sensor goes from about 0.18V to about 0.9V. Rear sensor signal is delayed and at a lower amplitude.
- Measured fuel pressure is 47 PSI. Rated expected range is 44 – 50 PSI.
- Front oxygen sensor switches from about 0.2V to about 0.9V
- Rear oxygen sensor is delayed in time and lower in amplitude

Turbocharging Normally Aspirated Engines on a Budget

Test data on engine before installation of turbocharger

Below are graphs of some of the data measured on the engine before the installation of the turbocharger, during a WOT acceleration run from 1000 RPM to 6700 RPM.

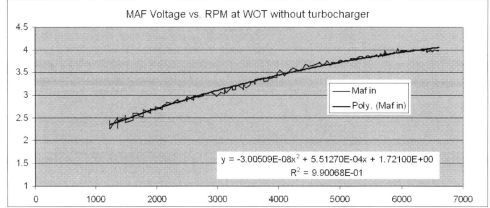

For the graphs below, the x-axis is time in seconds, from t = 15 seconds to 30 seconds of the data.

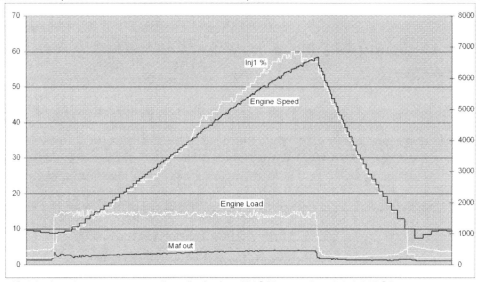

Average manifold absolute pressure (load) during WOT accel = 14.14 PSI.
- ✓ Injector duty cycle roughly tracks the RPM, as it should at a constant load.

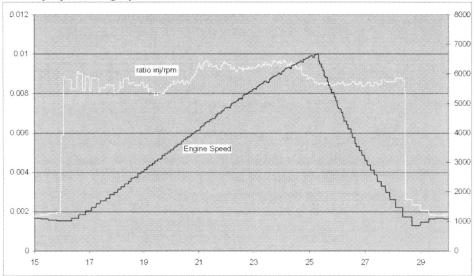

Average ratio inj/rpm = 0.008896 during WOT accel.

© 2012 Robert G. Wagoner
All rights reserved.

Turbocharging Normally Aspirated Engines on a Budget

MAF Sensor Modification

The MAF adapter I bought has an ID of 2.85". It requires building a sleeve to get it back to the original 2.52" ID. First I made a new tube for the MAF sensor to get it back to the original tube diameter of 2.52". The inner tube is made of a thin steel aerosol can 2.5" id. The tube is warped into a slight oval so the bottom edge of MAF sensor is at proper distance from bottom edge of tube, 0.62". A plastic insert in the original air box is shown in the pictures below. Originally I built my MAF tube without this insert, but the engine would not idle, and it would not run properly. I copied shape and position of the plastic insert into my MAF tube, built from the steel lid of a tuna can. Then the engine ran as good as it did with the original air box and intake tube. Based on the data from the AEM FIC, this new setup appears to flow a bit better at high RPM, as judged by the MAF data points being steadier at above 6000 RPM... possibly due to the elimination of the air box and intake tube.

Turbocharging Normally Aspirated Engines on a Budget

Initial Calculation of expected HP

The calculation of expected power is 199 HP with 8 PSI boost, as shown in the spreadsheet below.

	INPUTS	Source: "High Performance Fieros" by Robert Greg Wagoner
Outside Air Ambient Temp in deg F	60	
Altitude above Sea Level in ft	0	
Engine Volume in Liters	1.79	
Engine Compression Ratio	10.00	
Engine RPM for Maximum HP	6400	
Engine Volumetric Efficiency VE	0.895	Typical range is 0.70 to 0.85.
Engine Air to Fuel Ratio AFR	13.00	Should be rich. Enter 13.0 to 13.5.
Turbocharger Boost Pressure in psi (Gauge)	8.00	Set this to 0 if there is no turbocharger.
Turbocharger Compressor Efficiency CE	0.70	Find CE on the compressor map.
Intercooler Efficiency	0.75	Set this to 0 if there is no intercooler.
Intercooler Pressure Drop in psi	0.50	Set this to 0 if there is no intercooler.
X & Y VALUES FOR CE GRAPH		*formulas*
X = Air Flow in LBS / minute	20.1	B21*B22
Y = Turbocharger Boost Pressure Ratio PR	1.54	(B9+B19)/B19
OUTPUTS		
Outside Air Pressure in psia (Absolute)	14.7	14.7-B3*0.000494258
Boost Pressure Ratio after Intercooler	1.51	(B19+B9-B12)/B19
Air Flow in cubic ft / minute	263.2	0.0177516*B4*B6*B7*B10*B20/(B10+(1-B11)*(B16^0.283-1))
Air Density in LBS / cubic ft	0.076	2.7*B19/(B2+460)
Engine HP	199.4	298.2*B21*B22*B8*(1-1/B5^0.35)*(1-ABS(14.7-B8)/14.7)^2
Engine Torque in Ft-Lb	163.6	16500/pi()/B6*B23
Turbocharger Outlet Temperature in deg F	157.2	(B2+460)*(B16^0.283-1)/B10+B2
Intercooler Outlet Temperature in deg F	84.3	B25-B11*(B25-B2)

Calculations of Maximum HP based on Fuel Injector Size

Calculation based on existing fuel injector size and measured 47 PSI fuel pressure
- Fuel injector flow is 280cc/min x 4 = 1120cc/min at 100% duty cycle
- Maximum HP with these fuel injectors, using a conservative factor: 6cc/min / HP
  → 1120cc/min / (6cc/min / HP) = 186 HP at measured fuel pressure
- Initial goal was 199 HP at 6400 RPM with 8 PSI boost and with AFR=13.

These calculations give some confidence to go with the new plan to go with option 1), existing fuel injectors and existing MAF control. As soon as the car is running with boost, the exact boost pressure can be determined, based on measured AFR, as follows.

- Depending upon actual Air-to-Fuel Ratio (AFR) required to avoid detonation, the boost pressure may have to be decreased below 8 PSI if the AFR is very rich, or the boost pressure may be slightly higher than 8 PSI if the AFR is not so rich, such as AFR=13.5.

- Depending upon actual RPM where maximum HP is reached, the boost pressure may have to be decreased below 8 PSI if the maximum HP is reached at or above 6400 RPM, or the boost pressure may be slightly higher than 8 PSI if the maximum HP is reached below 6400 RPM.

© 2012 Robert G. Wagoner
All rights reserved.

Turbocharging Normally Aspirated Engines on a Budget

Installation of Turbocharger

The turbocharger and intercooler were installed next. The photo at the right is a view of the intercooler on the driver's side and the muffler on the passenger's side from the back of the car with the rear bumper cover and heat shields removed. The intercooler is supported by springs so it can move independent of the engine and independent of the body. The muffler is connected directly to the turbocharger so it moves with the engine.

The photo at the right is a view from the back of car with the rear bumper cover removed and the heat shields installed. The intercooler is isolated from the muffler by a heat shield, and fresh air flows to the intercooler from below the engine and out through vents in the rear of the bumper cover.

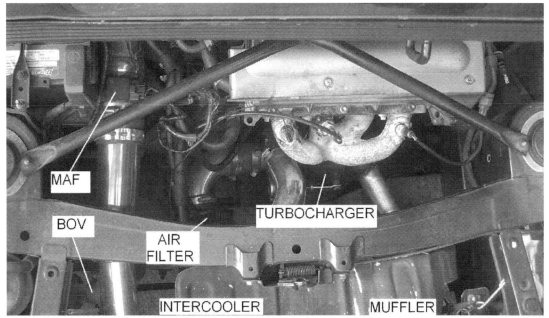

Top view showing locations of major components.

Turbocharging Normally Aspirated Engines on a Budget

The Turbocharger Itself

The turbocharger is a TB25 from a Saab 9000, part number TB2529, with a wastegate set for approximately 7 PSI boost pressure. This turbocharger has a small hole for a wastegate bypass, and internal wastegates have a reputation for "boost creep", and from the beginning I expected the total boost may rise above 7 PSI at high RPM. I planned to start low and conservative with the initial wastegate setting, and increase the boost pressure later if there is some margin to do that.

compressor wheel: 446335-0010   turbine wheel: 435922-0001

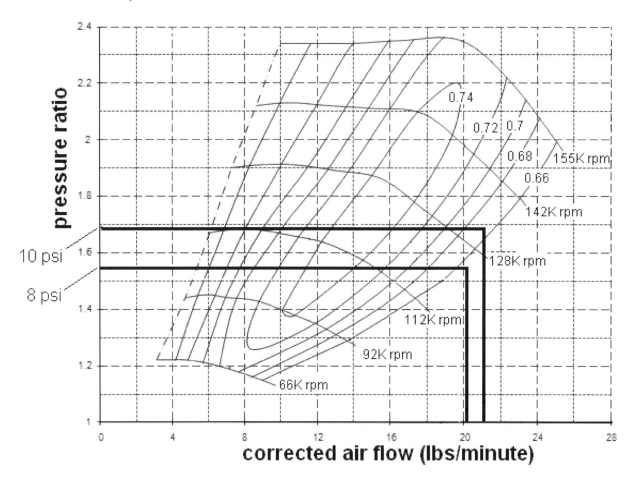

Water feed inlet and outlet are both ¼" FIP.
Oil feed is a stainless steel braided line 30" long, 1/8" FIP one end, M12x1.5 banjo bolt.

Turbocharging Normally Aspirated Engines on a Budget

For comparison to the graph on the next page, below is a graph of the engine load (absolute pressure in intake manifold) vs. RPM measured before the turbocharger was installed.

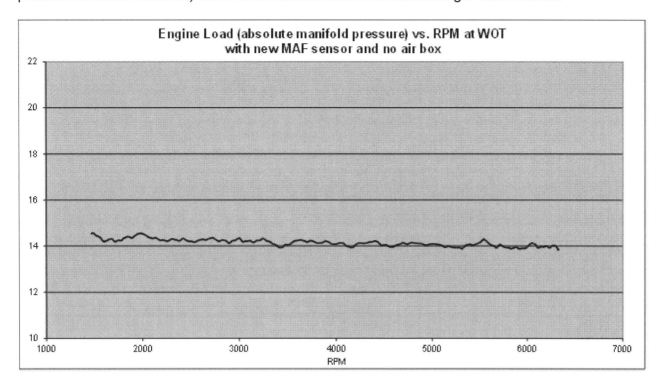

Initial Test of the Turbocharged Engine with No Compensation

I log data from the AEM FIC during Wide-Open-Throttle (WOT) runs in various gears and also during normal driving. Then I put the data into Excel to create graphs. The graphs are very helpful in understanding what is going on.

For an initial test run the AEM FIC has flat tables (no modifications to any parameters by the piggyback FIC). During the first test drive without stepping on the gas much it "feels" OK. Next test: WOT acceleration in first gear is a bit jerky, but it goes through first gear so fast I can't really figure out what I felt. Next I did a WOT acceleration in fourth gear, and clearly the engine bogs down at WOT somewhere in the range between 2500 – 3000 RPM. It is much more obvious in gears high enough to let it build boost at that RPM, such as fourth gear. In fourth gear the boost pressure on an external boost gauge is about 7 PSI at roughly 4000 RPM. At this point I decided I had taken enough data and to minimize risk I stopped to analyze the test data and give everything a visual inspection. I opened the hood and everything looked fine. Upon a quick inspection of the test data everything looked OK, so I went for a second test drive.

On a second test drive with all parameters still at zero (no compensation by the piggyback FIC) there was a considerable hiccup when going through the speed range approximately 2800 RPM, where it bogs down as if the fuel was shut off altogether, and another glitch at just below 2000 RPM, similar but less noticeable. Upon careful inspection of the fuel injector pulse width in this RPM range, it acts like the Toyota ECM switches from one mode of operation to another at 2800 RPM. This is not in the range where the MAF sensor runs out of range, and it is not in the range where the fuel injectors are running out of fuel. This is just a problem in the tuning of the Toyota ECM, where it was never expected to move this much fuel at this RPM. Clearly some adjustments will need to be made to the fuel, so the AEM FIC is absolutely necessary. Now I know there is no chance that the AEM FIC may be removed after the tuning is done.

© 2012 Robert G. Wagoner
All rights reserved.

Part 2: Tuning

In general the goal of tuning is to adjust the AFR and the ignition timing to get the car to run properly. To oversimplify the process, I make an adjustment to the AFR parameters, then I drive the car and measure (log) the AFR, and then I use Excel as a tool to analyze the recorded (logged) data. Then repeat over and over. There are many other software tools available for this tuning process, and many of them are designed specifically to make it easier to tune one type of engine.

There are books written devoted entirely to tuning. Hence it is not my goal to provide enough detail in this book to teach a person how to tune an engine. Similarly there are books written that teach the use of Excel, and similarly I cannot provide enough detail in this book to teach a person how to use Excel. The goal of this section is to give an overview of the tuning done on this particular engine, and to present the types of graphs and analyses I perform to tune an engine.

There were 55 revisions of the parameters to tune this setup. In the beginning the problems were obvious from the "feel" of the acceleration. At the end the differences could only be seen by looking at data and graphs, not by feel. A person could keep adjusting forever, but at some point a person has to quit and go on to the next project. To save space, only a few of the interim tuning steps are shown, particularly those that had noteworthy problems or improvements.

By revision 2 of tuning the fuel map, the glitch at 2800 RPM is less noticeable. However, there is a glitch at about 1800 RPM that is more difficult to tune out. At this point I was able to measure the boost pressure over the entire range of RPM. Below is a graph of engine load (absolute pressure in intake manifold) vs. RPM. Recall from the previous test without the turbocharger, the average manifold absolute pressure (load) during WOT was 14.14 PSI. The boost is approximately equal to the absolute pressure in the graph below minus 14.14 PSI.

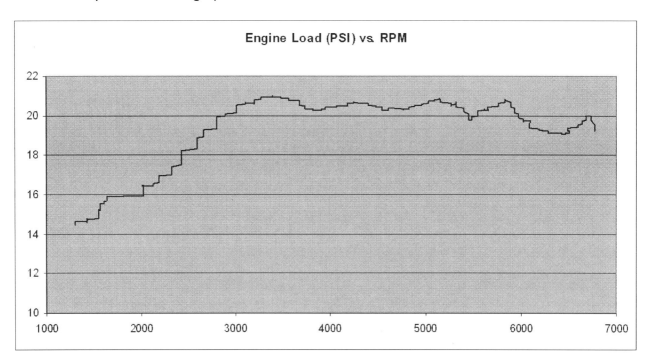

By revision 5, the car was running well enough to drive, but still had a lot of problems. For brevity I am skipping most of the revisions, showing only revision 5, revision 18, and the final one, revision 55. These three examples provide some graphs I find useful when tuning an engine. Comparing these three examples should give a sense of the types of changes made to tune this engine.

Turbocharging Normally Aspirated Engines on a Budget

AEM FIC Parameter set "FIC rev 5"

Timing map: no timing adjustments have been made, 0 everywhere.

MAF map: the MAF voltage has been reduced by the AEM FIC to follow the curve "mod MAF with boost" in the graph below. Curve "MAF no boost" is included for reference.

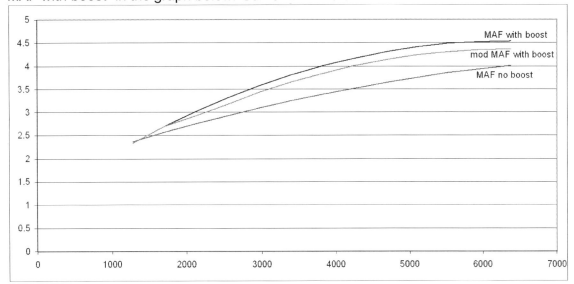

Fuel Injector map: (Percent additional injector pulse width vs absolute pressure vs. RPM)

Absolute PSI	0-2700 rpm	2850-4800 rpm	5100-5950 rpm	6375-8500 rpm
20 PSI	14.1	0	-5.5	1.6
17.5 PSI	11.7	0	-2.3	1.6
15 PSI	0	0	0	0
12.5 PSI	0	0	0	0

Driving feel: Feels OK generally. This pulls well through and above 2500 rpm range. Bogs down and has a glitch going through 1800-2000 rpm at WOT.

Logged test data: During a WOT acceleration run from 1500 RPM to 6800 RPM, at the right is a graph showing the ratio of injector duty cycle divided by rpm (light gray line and left vertical axis) and rpm (black line and right vertical axis). The horizontal axis is the time in seconds.

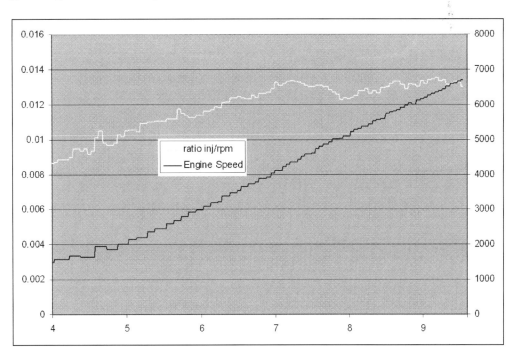

Turbocharging Normally Aspirated Engines on a Budget

I continued making slow changes, sometimes a few iterations each day, so that in roughly a week I had a good running car.

AEM FIC Parameter set "FIC rev 18"
Increased fuel slightly 2700-3400 rpm & 4800-5100.

Test data: Runs great. "Feels" great. No problems found during driving. However, according to the oxygen sensor, in 4th gear WOT run, it is lean in places: 1900 – 2800 rpm and above 3000 rpm. Next version add fuel in all these areas.

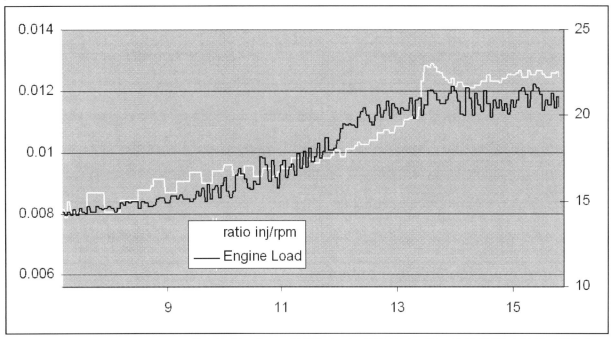

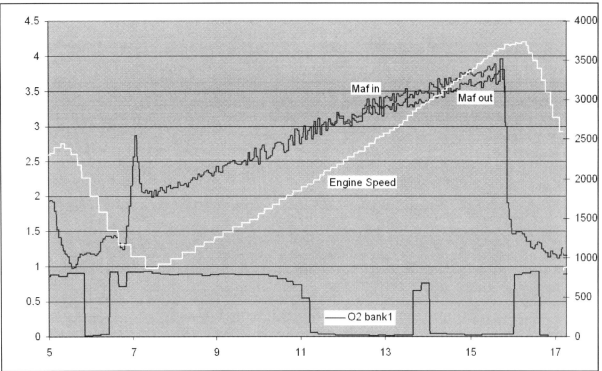

Turbocharging Normally Aspirated Engines on a Budget

For the next few months I continued to make very slight variations based on test data, iterations 19 through 54, not shown in here because they really add no valuable information except to demonstrate the amount of time it takes me to properly tune an engine.

AEM FIC Parameter set "FIC rev 55"

AEM FIC Parameter set "FIC rev 55" – fuel injector pulse width correction vs. RPM.

LOAD - MAP - PSIA	RPM: 0	1700	1900	2050	2300	2680	2900	3200	3400	4000	4350	4800	5100	5525	5950	6375	6800	7225	7650	8075	8500
40	0	0	0.8	15.6	19.5	23.4	7.8	7.8	6.3	3.1	0.8	1.6	1.6	-1.6	-2.3	-0.8	0	0	0	0	0
37.5	0	0	0.8	15.6	19.5	23.4	7.8	7.8	6.3	3.1	0.8	1.6	1.6	-1.6	-2.3	-0.8	0	0	0	0	0
35	0	0	0.8	15.6	19.5	23.4	7.8	7.8	6.3	3.1	0.8	1.6	1.6	-1.6	-2.3	-0.8	0	0	0	0	0
32.5	0	0	0.8	15.6	19.5	23.4	7.8	7.8	6.3	3.1	0.8	1.6	1.6	-1.6	-2.3	-0.8	0	0	0	0	0
30	0	0	0.8	15.6	19.5	23.4	7.8	7.8	6.3	3.1	0.8	1.6	1.6	-1.6	-2.3	-0.8	0	0	0	0	0
27.5	0	0	0.8	15.6	19.5	23.4	7.8	7.8	6.3	3.1	0.8	1.6	1.6	-1.6	-2.3	-0.8	0	0	0	0	0
25	0	0	0.8	15.6	19.5	23.4	7.8	7.8	6.3	3.1	0.8	1.6	1.6	-1.6	-2.3	-0.8	0	0	0	0	0
22.5	0	0	0.8	15.6	19.5	23.4	7.8	7.8	6.3	3.1	0.8	1.6	1.6	-1.6	-2.3	-0.8	0	0	0	0	0
20	0	0	0.8	15.6	19.5	23.4	7.8	7.8	6.3	3.1	0.8	1.6	1.6	-1.6	-2.3	-0.8	0	0	0	0	0
17.5	0	0	0.8	7.8	10.9	12.5	3.9	3.9	3.1	1.6	0	0.8	0.8	-0.8	-1.6	-0.8	0	0	0	0	0
15	0	0	0	0	0	0	0	0	0	0	0	0	0	0	0	0	0	0	0	0	0
12.5	0	0	0	0	0	0	0	0	0	0	0	0	0	0	0	0	0	0	0	0	0
10	0	0	0	0	0	0	0	0	0	0	0	0	0	0	0	0	0	0	0	0	0
7.5	0	0	0	0	0	0	0	0	0	0	0	0	0	0	0	0	0	0	0	0	0
5	0	0	0	0	0	0	0	0	0	0	0	0	0	0	0	0	0	0	0	0	0
2.5	0	0	0	0	0	0	0	0	0	0	0	0	0	0	0	0	0	0	0	0	0
0	0	0	0	0	0	0	0	0	0	0	0	0	0	0	0	0	0	0	0	0	0

AEM FIC Parameter set "FIC rev 55" – MAF sensor voltage correction vs. RPM.

LOAD - MAF - Volts	RPM: 0	425	850	1400	1700	2125	2550	2975	3400	3825	4250	4675	5100	5525	5950	6375	6800	7225	7650	8075	8500
5.5	0	0	0	0	0.8	0.8	-3.9	-3.9	-3.9	-3.9	-3.9	-3.9	-3.9	-3.9	-3.9	-3.9	-3.9	-3.9	-3.9	-3.9	-3.9
4.4	0	0	0	0	0.8	0.8	-3.9	-3.9	-3.9	-3.9	-3.9	-3.9	-3.9	-3.9	-3.9	-3.9	-3.9	-3.9	-3.9	-3.9	-3.9
4.3	0	0	0	0	0.8	0.8	-3.9	-3.9	-3.9	-3.9	-3.9	-3.9	-3.9	-3.9	-3.1	-2.3	-2.3	-2.3	-2.3	-2.3	-2.3
4.1	0	0	0	0	0.8	0.8	-3.9	-3.9	-3.9	-3.9	-3.9	-3.9	-3.1	-2.3	-1.6	-0.8	-0.8	-0.8	-0.8	-0.8	-0.8
4	0	0	0	0	0.8	0.8	-3.9	-3.9	-3.9	-3.9	-3.9	-3.1	-2.3	-1.6	-0.8	0	0	0	0	0	0
3.85	0	0	0	0	0.8	0.8	-3.9	-3.9	-3.9	-3.9	-3.1	-2.3	-1.6	0	0	0	0	0	0	0	0
3.75	0	0	0	0	0.8	0.8	-3.9	-3.9	-3.9	-3.1	-2.3	-1.6	0	0	0	0	0	0	0	0	0
3.65	0	0	0	0	0.8	0.8	-3.9	-3.9	-3.1	-2.3	-1.6	0	0	0	0	0	0	0	0	0	0
3.53	0	0	0	0	0.8	0.8	-3.9	-3.9	-2.3	-1.6	0	0	0	0	0	0	0	0	0	0	0
3.39	0	0	0	0	0.8	0.8	-3.9	-3.1	-1.6	0	0	0	0	0	0	0	0	0	0	0	0
3.25	0	0	0	0	0.8	0.8	-3.9	-1.6	0	0	0	0	0	0	0	0	0	0	0	0	0
3.1	0	0	0	0	0.8	0.8	-2.3	0	0	0	0	0	0	0	0	0	0	0	0	0	0
2.94	0	0	0	0	0.8	0.8	0	0	0	0	0	0	0	0	0	0	0	0	0	0	0
2.76	0	0	0	0	0.8	0	0	0	0	0	0	0	0	0	0	0	0	0	0	0	0
2.58	0	0	0	0	0	0	0	0	0	0	0	0	0	0	0	0	0	0	0	0	0
2.38	0	0	0	0	0	0	0	0	0	0	0	0	0	0	0	0	0	0	0	0	0
0	0	0	0	0	0	0	0	0	0	0	0	0	0	0	0	0	0	0	0	0	0

Turbocharging Normally Aspirated Engines on a Budget

AEM FIC Parameter set "FIC rev 55" – first set of test data, 2nd gear, WOT

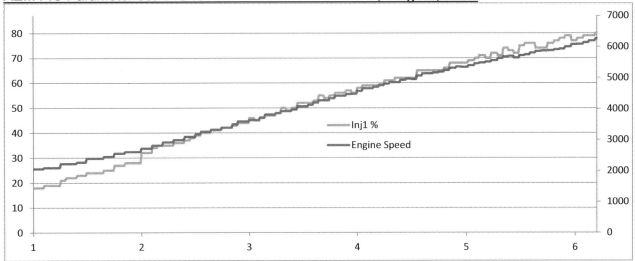

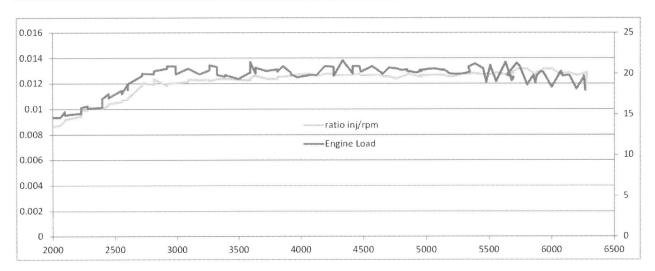

Runs great. "Feels" fine. Difficult to notice any bumps in performance by "feel", but they are sometimes still there at 2700 RPM. I like the way inj/rpm drifts from 0.012 at 3000 RPM to 0.013 as RPM increases. To me this makes sense, because at higher RPM more heat is generated, and extra fuel there will help cool the valves. This may also tend to improve fuel economy, since typical RPM is in the neighborhood of 3500 rpm.

Comparison of normally aspirated to turbocharged (averaged data from 3K – 6K rpm)
Rev 55 turbocharged data:
Average load = 20.21 PSI during WOT accel.
Average ratio inj/rpm = 0.012648 during WOT accel.
Time it took to go from 3K – 6K rpm in second gear = 3.52 seconds (slightly uphill)

Comparing to normally aspirated data:
Average load = 14.14 PSI during WOT accel.
Average ratio inj/rpm = 0.008896 during WOT accel.
Time it took to go from 3K – 6K rpm in second gear = 4.61 seconds (level road)

Desired ratio inj/rpm = 20.21/14.14*0.008896=0.012714. Close enough… for a while.

Turbocharging Normally Aspirated Engines on a Budget

AEM FIC Parameter set "FIC rev 55" – Scantool measurements

Below are Elmscan Scantool measurements of oxygen sensors bank 1 and bank 2 at 3000 rpm. Time scale is in minutes.

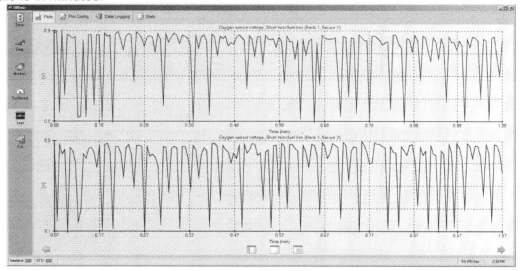

For comparison, below is data from the AEM FIC - measured oxygen sensors bank 1 and bank 2 at 3000 rpm. Bank 1 has faster sampling because it is measured by analog B input. Time scale is in seconds. Note, frequency is about 0.5 Hz. (rich = 0.8 volts, lean =0.1 volts).

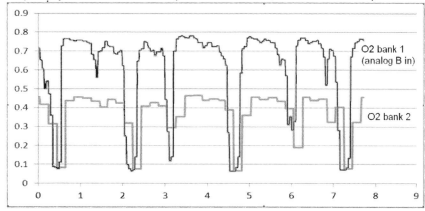

I took a series of data, in this case screen captures, using the Elmscan Scantool, to check long term and short term fuel trims. One example is below.

PID	Description	Value	Units
0x04	Calculated load value	14.51	%
0x07	Long term fuel % trim - Bank 1	2.34	%
0x06	Short term fuel % trim - Bank 1	-0.78	%
0x09	Long term fuel % trim - Bank 2	2.34	%
0x11	Absolute throttle position	11.76	%
0x10	Mass air flow rate	0.33	lb/min
0x0C	Engine RPM	952.25	RPM

© 2012 Robert G. Wagoner
All rights reserved.

Turbocharging Normally Aspirated Engines on a Budget

Below is a table made from the Elmscan data of long term and short term fuel trims. The more important one is the long term trim. Short term trim values tend to change a lot more and a lot faster. Checking long term trim is a good check to verify the tuning is OK. These will always drift around a bit depending on temperature, altitude, and various other factors. Trims of just a few percent are good. If the trims are a lot higher, the Air-to-Fuel Ratio based on the MAF map should probably be readjusted to reduce the fuel trims.

load	trim long 1	trim short 1	trim long 2	throttle	air flow	speed
%	%	%	%	%	lb/min	rpm
14.51	2.34	-0.78	2.34	11.76	0.33	952
13.33	2.34	-1.56	2.34	12.16	0.37	1005
63.14	1.56	0	0.78	69.41	3.31	1722
14.12	0.78	-3.91	-0.78	13.73	0.62	1751
13.73	0.78	-1.56	-0.78	14.12	0.68	1960
42.75	2.34	-7.81	0.78	20.78	2.41	2655
47.45	1.56	-7.03	0.78	20	2.87	2702
20.39	0	-0.78	0.78	17.25	1.59	3283

AEM FIC Parameter set "FIC rev 55" –test data, 4th gear, WOT

Injector duty cycle vs. RPM, 4th gear uphill

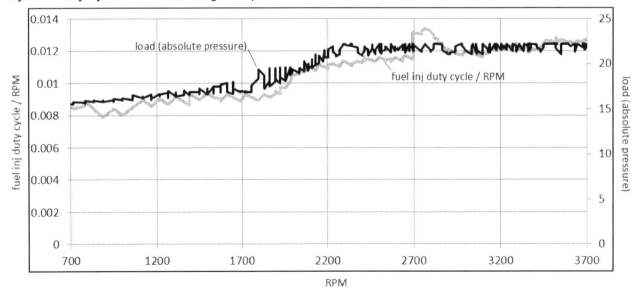

Turbocharging Normally Aspirated Engines on a Budget

Piggyback FIC Bypass Option - Impact on Availability

It is worth noting this particular choice of turbocharger setup, in particular using the original MAF size and original fuel injector size, as well as leaving the original ECM, while using the piggyback FIC, allows the possibility of easily going back to the original normally aspirated engine configuration. Simply bypass the piggyback FIC with the jumper they provide and open the wastegate to substantially reduce the boost pressure. Effectively the piggyback FIC is not absolutely necessary, and it could be considered a redundant system to improve the performance. This became very valuable to me when the MAP sensor built into the piggyback FIC failed. Temporarily I was able to continue driving the car by making the very changes I just described. I have to give credit to AEM for replacing the broken FIC for free under warranty, without any hassle.

Just for clarification, there is a big difference between availability and reliability. In this case adding the piggyback FIC has reduced the system reliability, but because it is effectively a redundant system, adding the piggyback FIC has NOT reduced the availability.

Finding Upper Limit of Boost Pressure

Here I go again trying to break stuff. After all the tuning described previously at 7-8 PSI boost, I changed the wastegate actuator to get 10 PSI boost as an experiment. By calculation I expected this to be the upper limit of the fuel injectors, and by test I found they were actually reaching 100% duty cycle at 6000 RPM, with the Air-To-Fuel Ratio (AFR) starting to increase above 13.5 at 6800 RPM. I do not recommend anyone else doing this. For a reliable setup with the stock fuel system, I suggest keeping the boost at 8.5 PSI maximum, which is the way I set my engine at the end.

I tested various wastegates actuators and various spring combinations. The graph below shows examples of the significant difference possible

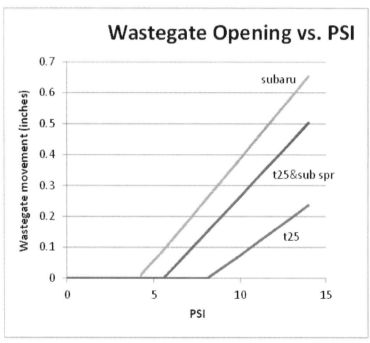

Top is the stock Subaru actuator
Middle is the T25 actuator with a spring from the Subaru actuator
Bottom is the stock T25 actuator

Turbocharging Normally Aspirated Engines on a Budget

The final wastegate actuator was set with a spring to result in approximately 9 PSI boost at the turbocharger, resulting in 8.5 PSI maximum boost measured in the intake manifold, as shown below. This keeps the fuel injector duty cycle below 100%, reaching 95% maximum at 6500 rpm at Wide Open Throttle (WOT) with AFR very close to 13.5.

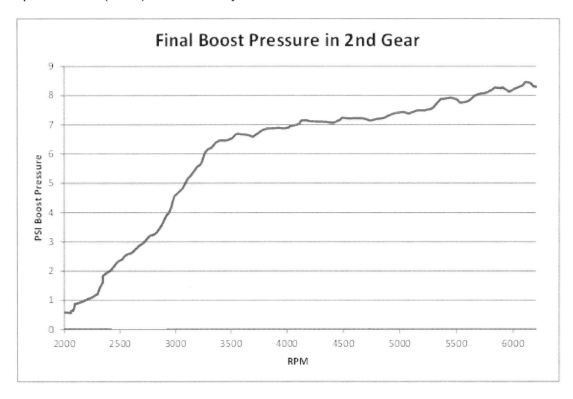

The calculated power is 207 HP with 9 PSI boost, as shown in the spreadsheet below.

	INPUTS	Source: "High Performance Fieros" by Robert Greg Wagoner
Outside Air Ambient Temp in deg F	60	
Altitude above Sea Level in ft	0	
Engine Volume in Liters	1.79	
Engine Compression Ratio	10.00	
Engine RPM for Maximum HP	6400	
Engine Volumetric Efficiency VE	0.895	Typical range is 0.70 to 0.85.
Engine Air to Fuel Ratio AFR	13.00	Should be rich. Enter 13.0 to 13.5.
Turbocharger Boost Pressure in psi (Gauge)	9.00	Set this to 0 if there is no turbocharger.
Turbocharger Compressor Efficiency CE	0.66	Find CE on the compressor map.
Intercooler Efficiency	0.75	Set this to 0 if there is no intercooler.
Intercooler Pressure Drop in psi	0.50	Set this to 0 if there is no intercooler.
X & Y VALUES FOR CE GRAPH		formulas
X = Air Flow in LBS / minute	20.8	B21*B22
Y = Turbocharger Boost Pressure Ratio PR	1.61	(B9+B19)/B19
OUTPUTS		
Outside Air Pressure in psia (Absolute)	14.7	14.7-B3*0.000494258
Boost Pressure Ratio after Intercooler	1.58	(B19+B9-B12)/B19
Air Flow in cubic ft / minute	272.9	0.0177516*B4*B6*B7*B10*B20/(B10+(1-B11)*(B16^0.283-1))
Air Density in LBS / cubic ft	0.076	2.7*B19/(B2+460)
Engine HP	206.8	298.2*B21*B22*B8*(1-1/B5^0.35)*(1-ABS(14.7-B8)/14.7)^2
Engine Torque in Ft-Lb	169.7	16500/pi()*B6*B23
Turbocharger Outlet Temperature in deg F	174.0	(B2+460)*(B16^0.283-1)/B10+B2
Intercooler Outlet Temperature in deg F	88.5	B25-B11*(B25-B2)

© 2012 Robert G. Wagoner
All rights reserved.

Turbocharging Normally Aspirated Engines on a Budget

Parts List and Costs

The table below is a list of the parts and costs for this turbocharger installation. Many of the parts were purchased on eBay, either new or used. These are sorted by cost, from most expensive to least expensive.

source	description	cost
ebay	AEM FIC part number 30-1910, new	$ 314.00
ebay	T25 turbocharger, new	$ 209.99
** local muffler shop	Exhaust down pipe with flange for turbocharger	$ 100.00
ebay	1ZZ turbo exhaust manifold, new	$ 85.00
ebay	Honda intercooler, used	$ 37.52
ebay	BOV, used	$ 26.50
Summit Racing	muffler, Walker part number 17887, size 2.25" in, 2.0" out, 20" long	$ 21.95
local auto parts store	oil return line, coolant hoses, & hose clamps	$ 16.68
ebay	stainless T-bolt clamps for intake, qty = 8	$ 16.15
local auto parts store	3" MAF adapter tube, Spectre part number 8705	$ 15.74
Summit Racing	2" dia tubing, mandrel bends, SUM-622001, for intake and exhaust	$ 14.95
local auto parts store	3" intake tube 90 degree bend, Spectre part number 8698	$ 12.59
local auto parts store	3" intake tube straight, Spectre part number 8718	$ 11.54
ebay	silicone 2" in-line coupling, qty = 2	$ 9.98
ebay	steel braided line 30" long, 1/8" FIP one end, M12x1.5 banjo bolt	$ 9.50
ebay	silicone 3" to 2.5" adapter	$ 8.00
ebay	3" Cone Air Filter, new	$ 7.99
Speedograph Richfield, Ltd	T-fitting for oil sensor, part number TP8, 1/8" BSPT to 1/8" x 27 NPT	$ 7.03
ebay	silicone 2" to 2.5" adapter	$ 5.99
ebay	small air filter for BOV vent to ambient	$ 3.88
local hardware store	1/4" Brass FIP fittings for coolant feed, qty = 2	$ 3.87
local auto parts store	3' hose 5/32" for coolant feed to turbocharger	$ 2.13
	grand total	$ 940.98

** This is an estimate for cost of custom exhaust work at a local shop. I make my own exhaust down pipes.

| Speedograph Richfield, Ltd | http://www.speedograph-richfield.com/html/oil_pipe_lines.html | |
| Summit Racing | http://www.summitracing.com/ | |

Breakage Cost

During this project, the cost of components broken was an extra $209, the cost of a turbocharger that failed after about 1000 miles (running too lean at high boost for a long trip: my fault). The extra parts for experiments, such as the Supra fuel injectors, were all sold later, and not included in this list of breakage cost.

© 2012 Robert G. Wagoner
All rights reserved.

> **Warning**
>
> *Modifications in this book may not be legal for street cars. Check local and state laws before making any changes to your car.*
>
> *Modifying your engine will void any factory warranties.*
>
> *The author does not recommend that anyone make any modifications to their car. The author is not responsible for any loss, damage, or injury caused by any modifications anyone makes to their car.*

Expect Breakage
- *I have had multiple major failures when modifying engines to increase power.*
- *I have learned to start a project like this with an expectation of major damage, and I feel lucky if that doesn't happen.*
- *I have done the most damage when I decided at the end to bump up the boost more and more until I "find the limit".*

Reliability
- *Minimize the number of modifications – every change reduces reliability.*
- *Start with a good-running, low-mileage engine.*
- *Ensure every modification is done carefully and correctly.*
- *Plan on a conservative power increase – don't push it.*

This book should be used in conjunction with a factory service manual, or a repair manual for your specific vehicle from GM, Chilton, or Haynes, which will include safety procedures, engine rebuilding information, torque specifications, cleaning, etc.

5 Turbocharging a 2L DOHC ECC Engine in a 1999 Dodge Neon

This is another example of turbocharging a normally aspirated engine on a budget. The initial goal was less than $1000, but it ended up closer to $1100 with the final addition of an ECM upgrade. Compared to the first two examples, this third example utilizes a much different fuel control, a mechanical fuel pressure controller to adjust fuel pressure based on boost pressure.

View of Turbocharged 2.0L DOHC from Driver's Side

The hardware installation is generally as described in chapter 3. These details will not be repeated in this chapter. There was no tuning required, with the exception of an ECM change. This chapter will focus on the differences unique to this particular turbocharger installation.

This Dual Overhead Camshaft (DOHC) engine, Chrysler engine code ECC (420A), is rated at 150 HP at 6500 RPM. It features 4 valves per cylinder, aluminum heads, and an aluminum intake manifold. The compression ratio is 9.6:1, and the actual displacement is 1.996 liters (~2 liters).

Fuel System Modification with 10:1 FMU

The initial plan for boost pressure is to set the boost pressure at 6 PSI at the turbocharger outlet, with 0.5 PSI pressure drop in the intercooler, leaving 5.5 PSI boost pressure in the intake manifold. Since this system has a MAP sensor, and it does not have a MAF sensor, and since I planned to keep the boost pressure relatively low (5.5 PSI at the intake manifold), this system will use a 10:1 rising rate boost fuel pressure regulator FMU (Fuel Management Unit) and an extra fuel pump. This would provide the means to add additional fuel without changing fuel injectors. The fuel pressure will be set so it will always run very rich with boost. With 5.5 PSI boost pressure, the 10:1 FMU will increase the fuel pressure by 55 PSI. The use of a 10:1 FMU is only a possibility with low boost pressures, because of the safety risk of additional fuel pressure. There will be no piggyback FIC, and it will use a stock ECM for control of the spark and fuel. It will be necessary to add a MAF voltage limiter as shown on page 36, to limit the voltage to 4.75VDC max, so it doesn't cut out or give error code at manifold pressures with boost. Simply put, when the ECM runs out of control at atmospheric pressure, as boost comes in, the fuel system takes over by itself to add more fuel. The ignition timing gets pinned at the timing which occurs when the MAP sensor hits its limit.

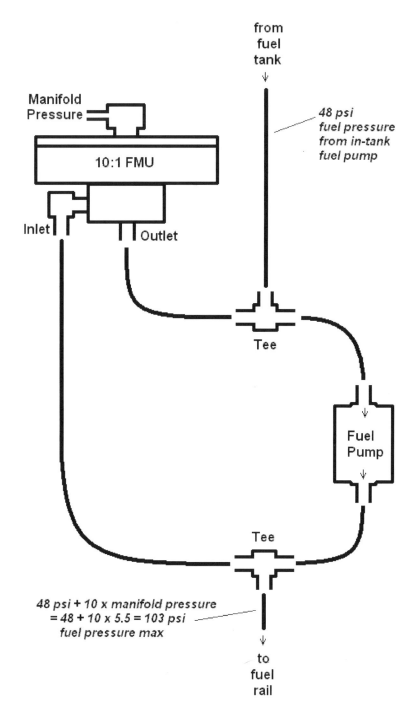

WARNING: High Fuel Pressure

This 10:1 FMU increases fuel pressure significantly. In this setup the fuel pressure can reach a maximum of 103 PSI. Increasing the fuel pressure is dangerous. The fuel pump control must remove power from all fuel pumps when the engine is not running, even if the ignition key is ON and the engine is stalled. Also the new fuel lines must be rated for the high pressure, and they must be made of material that is designed to work with fuel.

Turbocharging Normally Aspirated Engines on a Budget

Excel Spreadsheet Calculations of Airflow and Horsepower

The calculation of expected power is 192 HP with approximately 6 PSI boost at the turbocharger, resulting in 5.5 PSI maximum boost measured in the intake manifold, as shown in the spreadsheet below.

	INPUTS	Source: "High Performance Fieros" by Robert Greg Wagoner
Outside Air Ambient Temp in deg F	60	
Altitude above Sea Level in ft	0	
Engine Volume in Liters	2.00	
Engine Compression Ratio	9.60	
Engine RPM for Maximum HP	6500	
Engine Volumetric Efficiency VE	0.836	Typical range is 0.70 to 0.85.
Engine Air to Fuel Ratio AFR	13.00	Should be rich. Enter 13.0 to 13.5.
Turbocharger Boost Pressure in psi (Gauge)	6.00	Set this to 0 if there is no turbocharger.
Turbocharger Compressor Efficiency CE	0.72	Find CE on the compressor map.
Intercooler Efficiency	0.75	Set this to 0 if there is no intercooler.
Intercooler Pressure Drop in psi	0.50	Set this to 0 if there is no intercooler.
X & Y VALUES FOR CE GRAPH		*formulas*
X = Air Flow in LBS / minute	19.5	B21*B22
Y = Turbocharger Boost Pressure Ratio PR	1.41	(B9+B19)/B19
OUTPUTS		
Outside Air Pressure in psia (Absolute)	14.7	14.7-B3*0.000494258
Boost Pressure Ratio after Intercooler	1.37	(B19+B9-B12)/B19
Air Flow in cubic ft / minute	256.1	0.0177516*B4*B6*B7*B10*B20/(B10+(1-B11)*(B16^0.283-1))
Air Density in LBS / cubic ft	0.076	2.7*B19/(B2+460)
Engine HP	191.8	298.2*B21*B22/B8*(1-1/B5^0.35)*(1-ABS(14.7-B8)/14.7)^2
Engine Torque in Ft-Lb	154.9	16500/pi()/B6*B23
Turbocharger Outlet Temperature in deg F	133.5	(B2+460)*(B16^0.283-1)/B10+B2
Intercooler Outlet Temperature in deg F	78.4	B25-B11*(B25-B2)

Turbocharging Normally Aspirated Engines on a Budget

Turbocharger Selection

With a goal of 192 HP maximum on a 4 cylinder 2.0L engine, this engine could use a small turbocharger similar to the one selected for the MR2 Spyder in Chapter 4. However, the turbocharger I purchased for this engine was just slightly larger, a T3/T4, and it performed well. I chose the larger turbocharger because it did not have an internal wastegate, since an external wastegate was desired in this installation.

Hardware installation

The entire installation took only one weekend. First the exhaust manifold was replaced with a new one described below, and the turbocharger was mounted directly onto the exhaust manifold. More exhaust details are described in the section below. The oil feed and return were plumbed next.

The intercooler was mounted in front of the radiator. A universal intercooler piping kit was used to route the pressurized air from the turbocharger, located at the rear of the engine, under the engine in a gap between the engine and transmission, to the intercooler on the passenger's side of the car, then exiting the intercooler on the driver's side and up to the throttle body.

Exhaust / External Wastegate

The exhaust manifold was replaced with a cast iron turbocharger exhaust manifold (for a 420A engine) purchased on eBay. All new exhaust gaskets were used. An external 35mm wastegate was used to ensure good regulation of boost pressure with almost no boost creep, and because it makes it easy to set the boost pressure at the desired level. Using the weakest spring provided with the wastegate, it was possible to adjust the boost pressure to the desired 6 PSI. In actual use, this external wastegate did a very good job of regulating the boost pressure without boost creep.

A custom downpipe was made by modifying the original downpipe at the upper end to add the external wastegate and turbocharger exhaust flange. This minimized the amount of welding and the quantity of extra exhaust pipes needed. A flex pipe is included in the exhaust to allow the engine to twist without breaking the exhaust. The catalytic converter has been removed, and the oxygen sensor is located in the downpipe close to the turbocharger.

© 2012 Robert G. Wagoner
All rights reserved.

Turbocharging Normally Aspirated Engines on a Budget

Turbocharger System Block Diagram

The inlet air comes from the driver's side of the car behind and below the engine, where the air filter is attached directly to the turbocharger. From the air filter the air feeds into the turbocharger compressor inlet, then to the intercooler (located in front of the radiator), and finally into the throttle body, which is attached to the intake manifold, as shown below. No Blow-Off Valve (BOV) is used because this system has such low boost pressure.

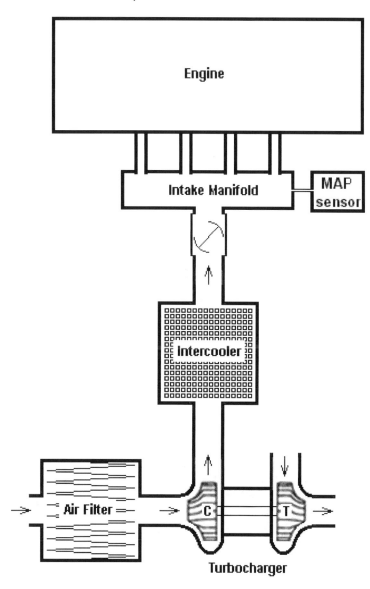

Ignition Timing

Initially I spent months trying to get this to work with the stock ECM. It kept cutting back the ignition timing, and would only allow it to run properly up to about 3 PSI boost pressure. The stock ECM was replaced with Mopar Performance 1-P5007038 ECM. This high performance ECM improved the ignition timing to allow proper operation up to the desired 6 PSI boost pressure. This high performance ECM also increased the RPM where the rev limiter cuts the engine, and it improved the "feel" of the rev limiter. The stock ECM dropped the RPM more when it hit the rev limiter before it allows the engine to resume operation, making it jerk more than this new high performance ECM.

© 2012 Robert G. Wagoner
All rights reserved.

Turbocharging Normally Aspirated Engines on a Budget

Cooling and Lubrication

Turbocharger water feed: the turbocharger coolant feed and return are like the ones in the previous chapters, tapped into the coolant loop through the throttle body. The details of the components used can be found in the parts list.

Turbocharger oil feed: the turbocharger oil feed and return are like the ones in the previous chapters, with a feed from the oil pressure sensor, and a return directly into the oil pan. The details of the components used can be found in the parts list.

Crankcase Ventilation

Because we eliminated the original intake filter and plumbing, the inlet for the PCV system was left unfiltered. Initially we installed a small air filter directly at the crankcase air inlet, exactly as we did on the Fiero. See page 34 for more details. This worked OK until we went to the track and did a severe test.

Crankcase Breather Problem

We found during Autocross racing when the car ran with boost pressure during high G turns, a little bit of the oil in the engine blew out of the crankcase breather air filter. A little oil pouring over the hot exhaust manifold makes a lot of smoke, which raises strong concerns! To correct this problem we built a small oil catch can that sits above the engine, between the crankcase air inlet and the small air filter, as shown below.

Turbocharging Normally Aspirated Engines on a Budget

FMU Reliability

The rising rate boost fuel pressure regulator FMU (Fuel Management Unit) seems to be an unreliable component. When they fail, the fuel pressure starts to creep up with no boost, resulting in the fuel pressure being too high particularly at an idle when the fuel flow to the engine is very low. This causes the engine to run too rich, idle rough, foul plugs, and make a lot of black smoke come from the exhaust. I tried three different FMU's from three different companies: one from Vortech, one from Kenne Bell, and one from OBX. Two failed during the three years I owned the car, and the third one failed after I sold the car. The person who bought the car from me reported to me the third one went bad after about a year of use, and he had to replace it again. Based on this experience I get the impression these are generally not very reliable.

Parts List and Costs

The table below is a list of the parts and costs for this turbocharger installation. Many of the parts were purchased on eBay, either new or used. These are sorted by cost, from most expensive to least expensive.

"Recommended" parts		
source	description	cost
ebay	Mopar Performance 1-P5007038 ECM, used	$ 176.00
ebay	T3/T04 turbocharger, new	$ 158.00
ebay	cast iron turbocharger exhaust manifold, 420A engine	$ 129.99
ebay	Walbro 255 LPH inline fuel pump, GSL392	$ 125.65
ebay	10:1 rising rate FMU	$ 101.30
** local muffler shop	Exhaust down pipe with flange for turbocharger	$ 100.00
ebay	universal intercooler piping kit	$ 89.99
ebay	external 35mm wastegate with flanges and gaskets	$ 61.99
ebay	Front-mount intercooler, new	$ 42.02
ebay	Polyester Synthetic Cone Air Filter & pipe, new	$ 36.77
local auto parts store	various exhaust gaskets	$ 31.58
local auto parts store	4 spark plugs	$ 10.53
local auto parts store	crankcase air breather	$ 10.44
local auto parts store	3' hose for fuel injected systems	$ 7.38
local auto parts store	8 hose clamps for fuel line	$ 6.64
local hardware store	two 1/4" Brass T	$ 4.18
	grand total	$ 1,092.46

** This is an estimate for cost of custom exhaust work at a local shop. I make my own exhaust down pipes.

Breakage Cost

During this project, the cost of broken components was an extra $224, not included in the parts list above. The major components that failed were two extra rising rate boost fuel pressure regulator FMU's. See the section above on Reliability for details.

Final Recommendation: Don't Use an FMU

I highly discourage using a rising rate boost Fuel Management Unit (FMU) based on the safety concerns I have from the high fuel pressure and also due to the unreliability of the FMU itself.

> **_Warning_**
> Modifications in this book may not be legal for street cars. Check local and state laws before making any changes to your car.
>
> Modifying your engine will void any factory warranties.
>
> The author does not recommend that anyone make any modifications to their car. The author is not responsible for any loss, damage, or injury caused by any modifications anyone makes to their car.

Expect Breakage
- *I have had multiple major failures when modifying engines to increase power.*
- *I have learned to start a project like this with an expectation of major damage, and I feel lucky if that doesn't happen.*
- *I have done the most damage when I decided at the end to bump up the boost more and more until I "find the limit".*

Reliability
- *Minimize the number of modifications – every change reduces reliability.*
- *Start with a good-running, low-mileage engine.*
- *Ensure every modification is done carefully and correctly.*
- *Plan on a conservative power increase – don't push it.*

This book should be used in conjunction with a factory service manual, or a repair manual for your specific vehicle from GM, Chilton, or Haynes, which will include safety procedures, engine rebuilding information, torque specifications, cleaning, etc.

Turbocharging Normally Aspirated Engines on a Budget

Turbocharging a 3.0L 2JZ-GE Engine in a 1995 Toyota Supra

This is another example of turbocharging a normally aspirated engine for less than $1000.

This is the easiest turbocharger installation of all in this book, because the fuel system, sensors, and ECM were designed by Toyota to come from the factory with enough margin to work properly with 8 PSI boost. The ECM control of Air-to-Fuel Ratio (AFR) is based on sensing of airflow through a MAF sensor, and as the boost pressure causes more airflow, the ECM is programmed to increase the pulse width of the fuel injectors to maintain the proper AFR. The extra manifold pressure does not cause a problem.

Turbocharging this engine is as simple as bolting on a turbocharger and doing the plumbing. It works with no tuning necessary. Anytime I start designing a turbocharger system for a new car, I always hope this is the case. Rarely am I this lucky.

Basic ideas and precautions about hardware installation are described in chapter 3. These details will not be repeated in this chapter. This chapter will focus on the differences peculiar to this particular turbocharger installation. Absolutely no tuning was required, with no changes to the fuel system or sensors.

The 2JZ-GE engine has Cylinder bore of 86 mm (3.39 in) and stroke of 86 mm (3.39 in), and with six in-line cylinders, the total displacement is 2.997 Liters. The compression ratio is 10.0:1. These engines are commonly converted from normally aspirated to turbocharged, and masses of information can be easily be found on the internet when searching "NA-T Supra". Numerous kits are available to make this change, ranging from over 300 HP just by bolting on the turbocharger, to over 400 HP with minor additional changes and a little more boost, and over 500 HP with more extensive modifications and more boost → all with high reliability of the stock bottom end and drivetrain. Recall the goal of this book is high reliability combined with low cost, and consequently the system I have chosen adds the fewest components with enough boost pressure to make a big difference in performance.

© 2012 Robert G. Wagoner
All rights reserved.

Turbocharging Normally Aspirated Engines on a Budget

Excel Spreadsheet Calculations of Airflow and Horsepower

The calculation of expected power is 306 HP with 8 PSI boost, as shown in the spreadsheet below.

	INPUTS	Source: "High Performance Fieros" by Robert Greg Wagoner
Outside Air Ambient Temp in deg F	60	
Altitude above Sea Level in ft	0	
Engine Volume in Liters	3.00	
Engine Compression Ratio	10.00	
Engine RPM for Maximum HP	5800	
Engine Volumetric Efficiency VE	0.905	Typical range is 0.70 to 0.85.
Engine Air to Fuel Ratio AFR	13.00	Should be rich. Enter 13.0 to 13.5.
Turbocharger Boost Pressure in psi (Gauge)	8.00	Set this to 0 if there is no turbocharger.
Turbocharger Compressor Efficiency CE	0.70	Find CE on the compressor map.
Intercooler Efficiency	0.75	Set this to 0 if there is no intercooler.
Intercooler Pressure Drop in psi	0.50	Set this to 0 if there is no intercooler.
X & Y VALUES FOR CE GRAPH		formulas
X = Air Flow in LBS / minute	30.8	B21*B22
Y = Turbocharger Boost Pressure Ratio PR	1.54	(B9+B19)/B19
OUTPUTS		
Outside Air Pressure in psia (Absolute)	14.7	14.7-B3*0.000494258
Boost Pressure Ratio after Intercooler	1.51	(B19+B9-B12)/B19
Air Flow in cubic ft / minute	403.3	0.0177516*B4*B6*B7*B10*B20/(B10+(1-B11)*(B16^0.283-1))
Air Density in LBS / cubic ft	0.076	2.7*B19/(B2+460)
Engine HP	305.6	298.2*B21*B22/B8*(1-1/B5^0.35)*(1-ABS(14.7-B8)/14.7)^2
Engine Torque in Ft-Lb	276.7	16500/pi()/B6*B23
Turbocharger Outlet Temperature in deg F	157.2	(B2+460)*(B16^0.283-1)/B10+B2
Intercooler Outlet Temperature in deg F	84.3	B25-B11*(B25-B2)

Turbocharger Selection

With a goal of 306 HP my initial choices were either a T60-1 or a T61 compressor. I found a kit on eBay that included a turbocharger, exhaust manifold, an external wastegate, and an oil line kit, all sized appropriately to work with this 2JZ-GE engine. The compressor curves are shown on the graph at the right, with the operating point from the Excel spreadsheet plotted on the graph.

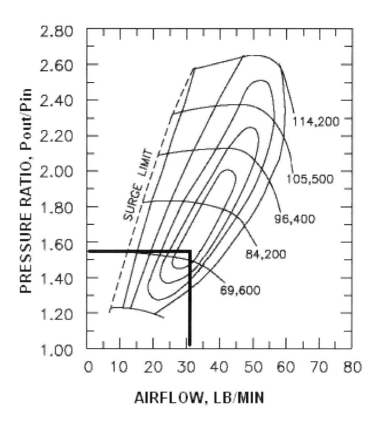

© 2012 Robert G. Wagoner
All rights reserved.

Turbocharging Normally Aspirated Engines on a Budget

Below is a photograph showing the intercooler behind the license plate, in front of the radiator.

Below is a photograph of the engine with the hood open, showing locations of major components.

Air Filter | MAF sensor | Pipe from Intercooler to Throttle Body | Turbocharger | Pipe from Turbocharger to Intercooler

© 2012 Robert G. Wagoner
All rights reserved.

Turbocharging Normally Aspirated Engines on a Budget

Turbocharger System Block Diagram

To turbocharge this engine, the components are connected as shown below. The inlet air is pulled from the front left side of the engine compartment where the new air filter is located. From the air filter the air feeds into the MAF sensor next, followed by the turbocharger compressor inlet, then to the intercooler, and finally to the throttle body, which is attached to the intake manifold, as shown below. No BOV was used because the MAF sensor location would have required complicated plumbing to route the BOV around the intercooler instead of directly to the outside air. Testing of the final system indicated this was a good choice, because no surge problem was found.

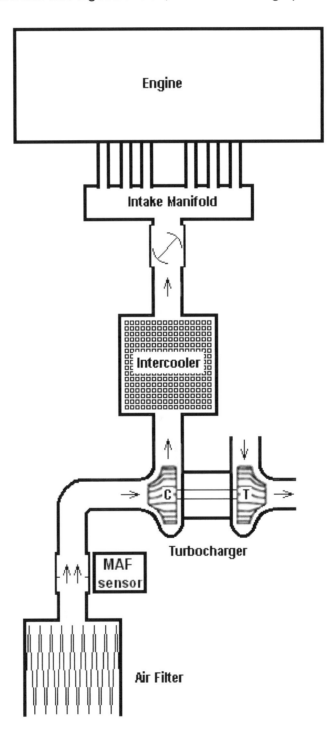

Turbocharging Normally Aspirated Engines on a Budget

Parts List and Costs

The table below is a list of the parts and costs for this turbocharger installation. Many of the parts were purchased on eBay, either new or used. These are sorted by cost, from most expensive to least expensive.

"Recommended" parts			
source	description	cost	
ebay	turbocharger, exhaust manifold, wastegate, oil lines, gaskets, new	$	453.98
ebay	3" intercooler piping with silicone rubber connectors & clamps	$	106.98
** local muffler shop	Exhaust down pipe for turbocharger	$	100.00
ebay	intercooler 31x12x3	$	76.00
ebay	3" silicone rubber connectors & clamps	$	28.78
ebay	3" to 2.5" silicone rubber reducer (qty 3)	$	17.97
ebay	new air filter	$	9.98
local auto parts store	oil return line (2 feet of 1/2" fuel compatible hose) & 2 hose clamps	$	9.01
local hardware store	oil return fitting (1/2" brass female hose barb to pipe thread)	$	7.86
	grand total	$	810.56

** This is an estimate for cost of custom exhaust work at a local shop. I make my own exhaust down pipes.

Breakage Cost

During this project, there was no breakage. I attribute this success to being very conservative in choice of the boost pressure, based on significant research before starting the project. To clarify this point, this engine and drivetrain is commonly pushed with higher boost to over 500 HP reliably.

© 2012 Robert G. Wagoner
All rights reserved.

> **Warning**
> Modifications in this book may not be legal for street cars. Check local and state laws before making any changes to your car.
>
> Modifying your engine will void any factory warranties.
>
> The author does not recommend that anyone make any modifications to their car. The author is not responsible for any loss, damage, or injury caused by any modifications anyone makes to their car.

Expect Breakage
- *I have had multiple major failures when modifying engines to increase power.*
- *I have learned to start a project like this with an expectation of major damage, and I feel lucky if that doesn't happen.*
- *I have done the most damage when I decided at the end to bump up the boost more and more until I "find the limit".*

Reliability
- *Minimize the number of modifications – every change reduces reliability.*
- *Start with a good-running, low-mileage engine.*
- *Ensure every modification is done carefully and correctly.*
- *Plan on a conservative power increase – don't push it.*

> *This book should be used in conjunction with a factory service manual, or a repair manual for your specific vehicle from GM, Chilton, or Haynes, which will include safety procedures, engine rebuilding information, torque specifications, cleaning, etc.*

7 Turbocharging a 1.8L Engine in a 1997 Mazda Miata

This is another example of turbocharging a normally aspirated engine for approximately $1000. The hardware installation is generally as described in chapter 3, and tuning methods are described in chapter 4. These details will not be repeated in this chapter. This chapter will focus on the differences peculiar to this particular turbocharger installation.

Turbocharging Normally Aspirated Engines on a Budget

This engine, designated as BP-ZE, comes normally aspirated from the factory with ratings of 133 HP at 6000 rpm and 122 ft-lb of torque at 4000 rpm. It has an aluminum alloy head with Dual Overhead Camshaft (DOHC) and four valves per cylinder. For turbocharging considerations, it has a compression ratio of 9.0:1 and electronic fuel injection based on a vane-type mass air flow meter.

System Block Diagram

To turbocharge this 1.8L engine, the air filter is connected directly to the MAF sensor inlet, and next to the turbocharger compressor inlet. The compressed air flows to the intercooler (located in front of the engine), and then to the throttle body. A Blow-Off Valve (BOV) is located in a way to bypass the turbocharger. Since the MAF sensor is located before the BOV, the air "leak" caused by the BOV would have an adverse influence on the ability of the MAF sensor to control the Air-Fuel Ratio, unless the BOV outlet is routed back into the intake piping after the MAF sensor, as shown below.

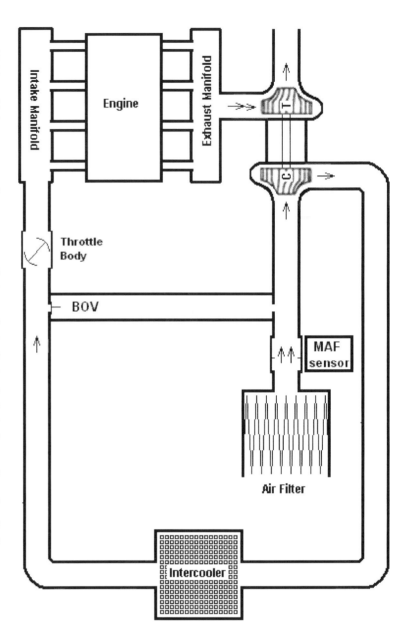

Head porting

The heads on this engine were ported and polished, as shown in the photographs below. It takes a LOT of time and effort to get them looking this nice. Kudos to Seth for this excellent work.

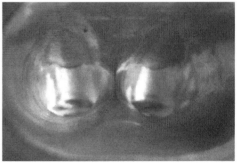

Turbocharging Normally Aspirated Engines on a Budget

The location of the intercooler at the front of the car can be seen in the photograph below.

A close-up photograph of the turbocharger exhaust configuration is shown below.

Turbocharging Normally Aspirated Engines on a Budget

Excel Spreadsheet Calculations of Airflow and Horsepower

The calculation of expected power is 199 HP with 10 PSI boost, as shown in the spreadsheet below. Dynamometer measurements of this car show the actual peak is 205 HP, which would be predicted by the Excel spreadsheet at either slightly higher boost pressure or slightly higher RPM.

	INPUTS	Source: "High Performance Fieros" by Robert Greg Wagoner
Outside Air Ambient Temp in deg F	60	
Altitude above Sea Level in ft	0	
Engine Volume in Liters	1.80	
Engine Compression Ratio	9.00	
Engine RPM for Maximum HP	6000	
Engine Volumetric Efficiency VE	0.910	Typical range is 0.70 to 0.85.
Engine Air to Fuel Ratio AFR	13.00	Should be rich. Enter 13.0 to 13.5.
Turbocharger Boost Pressure in psi (Gauge)	10.00	Set this to 0 if there is no turbocharger.
Turbocharger Compressor Efficiency CE	0.66	Find CE on the compressor map.
Intercooler Efficiency	0.75	Set this to 0 if there is no intercooler.
Intercooler Pressure Drop in psi	0.50	Set this to 0 if there is no intercooler.
X & Y VALUES FOR CE GRAPH		*formulas*
X = Air Flow in LBS / minute	20.7	B21*B22
Y = Turbocharger Boost Pressure Ratio PR	1.68	(B9+B19)/B19
OUTPUTS		
Outside Air Pressure in psia (Absolute)	14.7	14.7-B3*0.000494258
Boost Pressure Ratio after Intercooler	1.65	(B19+B9-B12)/B19
Air Flow in cubic ft / minute	271.0	0.0177516*B4*B6*B7*B10*B20/(B10+(1-B11)*(B16^0.283-1))
Air Density in LBS / cubic ft	0.076	2.7*B19/(B2+460)
Engine HP	199.1	298.2*B21*B22/B8*(1-1/B5^0.35)*(1-ABS(14.7-B8)/14.7)^2
Engine Torque in Ft-Lb	174.3	16500/pi()/B6*B23
Turbocharger Outlet Temperature in deg F	184.6	(B2+460)*(B16^0.283-1)/B10+B2
Intercooler Outlet Temperature in deg F	91.2	B25-B11*(B25-B2)

The compressor curves for this T28 turbocharger, along with this operating point, are shown below.

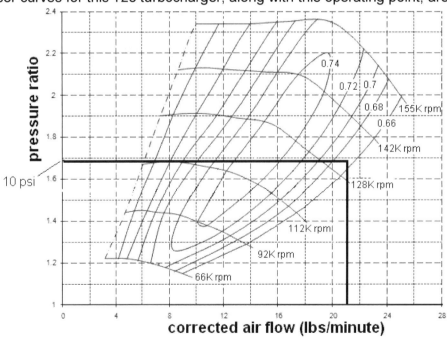

© 2012 Robert G. Wagoner
All rights reserved.

Turbocharging Normally Aspirated Engines on a Budget

Replacing a 1.6L Miata engine with a 1.8L Miata engine

In this particular situation, the car started with the 1.6L engine, which was replaced with a 1.8L engine. In order to allow the new engine to work with the original ECM, it is necessary to keep the 1.6 throttle body. Adapting this throttle body to the 1.8L intake manifold is the biggest part of the job. *Flyin Miata* makes an adaptor plate to do this for just under $60. This is not included in the parts list below, because it would not be necessary if the car started with the 1.8L engine.

Parts List and Costs

The table below is a list of the parts and costs for this turbocharger installation. Many of the parts were purchased on eBay, either new or used. These are sorted by cost, from most expensive to least expensive. In order to keep the cost of this installation at a minimum, all expensive parts were purchased used, with the exception of the exhaust manifold, which was purchased new. Not all costs of every component used for this particular installation were tracked completely, and the prices below reflect values of these components on eBay.

"Recommended" parts		
source	description	cost
ebay	turbocharger, no-name from China	$ 350.00
ebay	piggy back Greddy Emanage ECU, used	$ 232.49
**(1) ebay	OBX cast manifold, new	$ 119.00
ebay	2" intercooler piping with silicone rubber connectors & clamps, new	$ 106.98
**(2) local muffler shop	Exhaust down pipe for turbocharger	$ 100.00
ebay	27" bar and plate front mount intercooler, new	$ 74.98
ebay	HKS blow off valve, new	$ 64.99
ebay	red top 550cc rx7 injectors, used	$ 50.45
ebay	rx7 high flow MAF sensor, used	$ 42.77
local hardware store	grinding & sanding wheels for porting / polishing the head	$ 27.76
ebay	oil feed kit	$ 18.90
ebay	air filter, new	$ 9.98
local auto parts store	oil return line (2 feet of 1/2" fuel compatible hose) & 2 hose clamps	$ 9.01
local hardware store	oil return fitting (1/2" brass female hose barb to pipe thread)	$ 7.86
	grand total	$ 1,215.17

**(1) The exhaust manifold used in the photographs was actually made by flying miata, which costs $395.
**(2) This is an estimate for custom exhaust work at a local shop. We avoid this cost by doing our own welding.

Breakage Cost

During this project, the cost of broken components not counted in the list above was at least an extra $1600, but who's counting? The damaged major components are listed below.

- The differential failed first. Replaced the broken 4.3 with a stronger 4.1 Torsen LSD ($600).
- Next the transmission failed. Replaced the broken 5-speed with a 6-speed ($650).
- Also a turbocharger went bad (bearings damaged) and was replaced ($350 from China)

Turbocharging Normally Aspirated Engines on a Budget

> **_Warning_**
> _Modifications in this book may not be legal for street cars. Check local and state laws before making any changes to your car._
>
> _Modifying your engine will void any factory warranties._
>
> _The author does not recommend that anyone make any modifications to their car. The author is not responsible for any loss, damage, or injury caused by any modifications anyone makes to their car._

Expect Breakage
- *I have had multiple major failures when modifying engines to increase power.*
- *I have learned to start a project like this with an expectation of major damage, and I feel lucky if that doesn't happen.*
- *I have done the most damage when I decided at the end to bump up the boost more and more until I "find the limit".*

Reliability
- *Minimize the number of modifications – every change reduces reliability.*
- *Start with a good-running, low-mileage engine.*
- *Ensure every modification is done carefully and correctly.*
- *Plan on a conservative power increase – don't push it.*

> *This book should be used in conjunction with a factory service manual, or a repair manual for your specific vehicle from GM, Chilton, or Haynes, which will include safety procedures, engine rebuilding information, torque specifications, cleaning, etc.*

Turbocharging Normally Aspirated Engines on a Budget

8 Turbocharging a 2.8L V-6 Engine in a 1987 Pontiac Fiero

This is another example of turbocharging a normally aspirated engine, in this case a variation based on the 3.4L Fiero in Chapter 3, with a goal of even lower cost, dropping the price to of the entire project to roughly $500. Most of the information in chapter 3 is applicable to this example, and only the differences will be described here to avoid a lot of repetition. The major differences are: this Fiero has the original 2.8L engine, a smaller turbocharger, no intercooler, and no BOV. Because of the problems we experienced adding an intercooler in the trunk of the Fiero, causing intercooler piping to pull apart as described in Chapter 3, this is a more reliable setup without an intercooler. This turbocharger is operated at 8 PSI boost, compared to the 11 PSI boost of the 3.4L in chapter 3. With a smaller engine and lower boost, this 2.8L engine has much less power than the 3.4L.

Turbocharger: The turbocharger location was chosen exactly as described in Chapter 3 on page 28. We were trying to keep this project low budget if possible, so in order to save money we decided to purchase a used turbocharger and rebuild it. We found many small used turbochargers on eBay for a relatively low price. We wanted a Garrett T3 with a "45" size compressor. This limited our options greatly. After searching for a few weeks, we found one on eBay that fit our requirements. We placed a high bid on it, and ended up getting it for a reasonable price. When it arrived we found it was as described, except it had a small crack in the exhaust side near the wastegate, a typical problem with these old Garrett T3 turbochargers, but one which does not have any significant effect on the performance. We rebuilt the turbocharger before installing it. For details see "Rebuilding a Turbocharger" in Appendix F. Looking back, with my experience now, I recommend purchasing a new turbocharger instead of buying used and rebuilding it.

Turbocharging Normally Aspirated Engines on a Budget

Turbocharged 2.8L Fiero with Engine Vent Cover Removed.

Turbocharged 2.8L Fiero with Engine Vent Cover Installed.

Turbocharging Normally Aspirated Engines on a Budget

Excel spreadsheet calculations of the turbocharger compressor.

	INPUTS	*Source: "High Performance Fieros" by Robert Greg Wagoner*
Outside Air Ambient Temp in deg F	60	
Altitude above Sea Level in ft	0	
Engine Volume in Liters	2.80	
Engine Compression Ratio	8.90	
Engine RPM for Maximum HP	5000	
Engine Volumetric Efficiency VE	0.714	Typical range is 0.70 to 0.85.
Engine Air to Fuel Ratio AFR	13.00	Should be rich. Enter 13.0 to 13.5.
Turbocharger Boost Pressure in psi (Gauge)	8.00	Set this to 0 if there is no turbocharger.
Turbocharger Compressor Efficiency CE	0.65	Find CE on the compressor map.
Intercooler Efficiency	0.00	Set this to 0 if there is no intercooler.
Intercooler Pressure Drop in psi	0.00	Set this to 0 if there is no intercooler.
X & Y VALUES FOR CE GRAPH		*formulas*
X = Air Flow in LBS / minute	17.4	B21*B22
Y = Turbocharger Boost Pressure Ratio PR	1.54	(B9+B19)/B19
OUTPUTS		
Outside Air Pressure in psia (Absolute)	14.7	14.7-B3*0.000494258
Boost Pressure Ratio after Intercooler	1.54	(B19+B9-B12)/B19
Air Flow in cubic ft / minute	228.1	0.0177516*B4*B6*B7*B10*B20/(B10+(1-B11)*(B16^0.283-1))
Air Density in LBS / cubic ft	0.076	2.7*B19/(B2+460)
Engine HP	167.0	298.2*B21*B22/B8*(1-1/B5^0.35)*(1-ABS(14.7-B8)/14.7)^2
Engine Torque in Ft-Lb	175.4	16500/pi()/B6*B23
Turbocharger Outlet Temperature in deg F	164.7	(B2+460)*(B16^0.283-1)/B10+B2
Intercooler Outlet Temperature in deg F	164.7	B25-B11*(B25-B2)

Excel spreadsheet with parameters for a Garrett T03-"45" operated at 8 PSI without an intercooler.

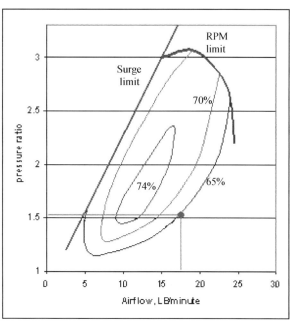

Compressor curves for Garrett T03-"45"

© 2012 Robert G. Wagoner
All rights reserved.

Turbocharging Normally Aspirated Engines on a Budget

Turbocharger System Block Diagram

To turbocharge this 2.8L V6 engine in a 1987 Pontiac Fiero, the inlet air is pulled from the driver's side of the car where the air filter is located. From the air filter the air will feed into the turbocharger compressor inlet, and then into the throttle body, which is attached to the intake manifold, as shown below. This system will require changing from the stock MAP sensor to a 2 BAR MAP sensor, as described previously on page 35.

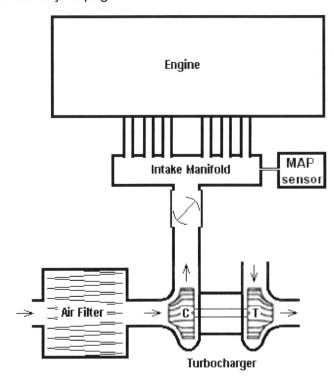

Fuel Injectors: This engine comes from the factory with 15 LB/hr fuel injectors, which cannot move enough fuel to add the turbocharger. We changed to 19LB/hr fuel injectors, which operate at 87.9% maximum duty cycle at 167 HP, as calculated below.

$$\text{Fuel injector duty cycle} = 167 \times 0.6 / 6 / 19 = 0.879 = 87.9 \% \text{ duty cycle}$$

Fuel system: Everything on the fuel system is stock, including the fuel pump. The stock Fiero fuel pump has plenty of margin at this HP level, based on calculation in chapter 3.

Engine, Clutch, and Transmission: This particular engine is stock, with stock heads, stock camshaft, stock oil pump, and all other stock internal parts. No change was made to the engine except the fuel injectors and MAP sensor, as described previously. The stock clutch has the margin to handle the extra torque of the 2.8L engine with 7 PSI boost pressure.

Engine Support: We upgraded the engine strut support motor mount on the passenger's side of the engine with a Prothane kit p/n 7-501. All other motor and transmission mounts are stock.

Oil Lubrication System: The oil feed for the turbocharger is set up exactly as described in Chapter 3 on page 34.

Exhaust was modified similar to the 3.4L Fiero engine, except the turbocharger is twisted slightly more in this case, to align the turbocharger outlet more directly with the throttle body.

Turbocharging Normally Aspirated Engines on a Budget

Cooling and Lubrication

Turbocharger water feed: the turbocharger coolant feed and return are tapped into the coolant loop through the throttle body just like the one in Chapter 3.

Turbocharger oil feed: the turbocharger oil feed and return are like the one in Chapter 3, with a feed from the oil pressure sensor, and a return directly into the oil pan.

Crankcase Ventilation

Because we eliminated the original intake filter and plumbing, the inlet for the PCV system was left unfiltered. We installed a small air filter directly at the crankcase air inlet, exactly as we did on the 3.4L Fiero engine. See page 34 for more details.

Parts List and Costs

The table below is a list of the parts and costs for this turbocharger installation. Many of the parts were purchased on eBay, either new or used.

"Recommended" parts		
source	description	cost
ebay	T3-"45" turbocharger, new no-name brand	$ 228.50
local auto parts store	2 BAR MAP sensor	$ 64.57
ebay	6 of 19 LB fuel injectors, used	$ 45.20
ebay	Polyester Synthetic Cone Air Filter & pipe, new	$ 36.77
ebay	Prothane Kit 7-501 for Engine Torque Strut	$ 22.97
local muffler shop	misc exhaust pipes	$ 20.00
local muffler shop	Exhaust pipe with flange for turbocharger	$ 15.67
ebay	two silicone joiner connectors	$ 12.90
local auto parts store	2 of 1/4"F to 1/8M 90deg, brake line 1/4" x 60"	$ 12.28
local auto parts store	6 of NGK UR6 spark plugs	$ 10.53
local auto parts store	red air breather	$ 10.44
local hardware store	transmission mount	$ 9.39
local hardware store	pipe, tape, & hose clamps	$ 6.96
local auto parts store	3' hose 5/32"	$ 2.13
local auto parts store	brass 5/16" hose to 3/8"FUE	$ 1.72
local hardware store	1/4" Brass T and 1/4" brass nipple	$ 3.87
	grand total	$ 503.90

Deck Lid Supports were changed from the original springs to pressurized gas struts. However, this change was not absolutely necessary, but more for aesthetic value. These are not included in the "recommended parts" list.

Breakage Cost

During this project, there was no breakage. I attribute this success to being very conservative in choice of the boost pressure, and because this was based on a previous project with significant breakage cost, where I learned how to get it right this time.

Turbocharging Normally Aspirated Engines on a Budget

Tuning

The table below shows the final results of Volumetric Efficiency (VE) and spark advance. Each installation is a bit different, and these tables of parameters would be a good starting point.

MAP voltage	0.511	1.040	1.569	2.099	2.628	3.157	3.686	4.215	4.745
BAR	0.4	0.6	0.8	1.0	1.2	1.4	1.6	1.8	2.0
800 RPM	6.3	18	20.7	27.3	32.4	37.1	42.2	47.3	52.3
1200 RPM	9.4	21.1	23.4	29.7	34.8	39.8	44.9	49.6	54.7
1600 RPM	14.5	24.2	26.6	32	37.1	42.2	47.3	52.3	57
2000 RPM	19.5	27.3	30.1	34.8	39.8	44.5	49.6	54.7	59.8
2400 RPM	21.9	30.5	33.6	37.1	42.2	47.3	52	57	62.1
2800 RPM	23.8	33.2	35.9	39.5	44.5	49.6	54.7	59.4	64.5
3200 RPM	27	35.2	38.3	42.2	46.9	52	57	62.1	67.2
3600 RPM	29.7	36.3	39.8	44.5	49.6	54.3	59.4	64.5	69.5
4000 RPM	33.2	38.7	41.4	46.9	52	57	61.7	66.8	71.9

VE table for small turbocharger with 2 BAR MAP sensor and 19 LB fuel injectors, VE multiplier = 224

MAP voltage	0.511	0.776	1.040	1.305	1.569	1.834	2.099	2.363	2.628	2.892	3.157	3.422	3.686	3.951	4.215	4.480	4.745
BAR	0.4	0.5	0.6	0.7	0.8	0.9	1	1.1	1.2	1.3	1.4	1.5	1.6	1.7	1.8	1.9	2
600 RPM	28.1	28.1	28.1	28.1	28.1	13.4	5.6	2.5	0	0	0	0	0	0	0	0	0
800 RPM	32	32	32	32	28.8	20	13	9.8	7	3.9	1.1	0	0	0	0	0	0
1000 RPM	35.9	35.9	35.9	35.9	28.8	20	13	9.8	7	3.9	1.1	0	0	0	0	0	0
1200 RPM	40.1	40.1	40.1	35.9	33	26	16.9	14.1	10.9	7.7	4.9	1.8	0	0	0	0	0
1400 RPM	41.8	41.8	40.1	35.2	33	27.1	17.9	14.8	12	8.8	6	2.8	0	0	0	0	0
1600 RPM	46.1	46.1	40.1	35.9	34.1	27.1	19	16.2	13	9.8	7	3.9	1.1	0	0	0	0
2000 RPM	46.1	46.1	41.8	35.9	34.1	28.8	23.9	20.7	17.9	14.8	12	8.8	6	2.8	0	0	0
2400 RPM	46.1	46.1	46.1	40.1	35.2	29.9	26	22.9	20	16.9	14.1	10.9	8.1	4.9	2.1	0	0
2800 RPM	48.2	48.2	46.1	40.1	36.9	29.9	27.1	24.3	21.1	17.9	15.1	12	9.1	6	3.2	0	0
3200 RPM	45	45	41.8	40.1	36.9	33	28.8	25.7	22.9	19.7	16.9	13.7	10.9	7.7	4.9	1.8	0
3600 RPM	41.8	41.8	40.1	39	35.2	28.1	28.1	25.0	22.1	19	16.2	13	10.2	7	4.2	1.1	0
4000 RPM	45	45	45	39	35.2	29.9	28.1	25.0	22.1	19	16.2	13	10.2	7	4.2	1.1	0
4400 RPM	48.2	48.2	46.1	39	35.9	33	29.9	27.1	23.9	20.7	17.9	14.8	12	8.8	6	2.8	0
4800 RPM	49.9	49.9	46.1	39	35.9	33	29.9	27.1	23.9	20.7	17.9	14.8	12	8.8	6	2.8	0

Spark Advance table for small turbocharger and 2 BAR MAP sensor

Turbocharging Normally Aspirated Engines on a Budget

Below are example photographs of a very similar turbocharger installation into a 2.8L V6 in a 1986 Fiero SE, except the air filter is located at the top instead of routed down toward the bottom.

> **Warning**
> Modifications in this book may not be legal for street cars. Check local and state laws before making any changes to your car.
>
> Modifying your engine will void any factory warranties.
>
> The author does not recommend that anyone make any modifications to their car. The author is not responsible for any loss, damage, or injury caused by any modifications anyone makes to their car.

Expect Breakage
- I have had multiple major failures when modifying engines to increase power.
- I have learned to start a project like this with an expectation of major damage, and I feel lucky if that doesn't happen.
- I have done the most damage when I decided at the end to bump up the boost more and more until I "find the limit".

Reliability
- Minimize the number of modifications – every change reduces reliability.
- Start with a good-running, low-mileage engine.
- Ensure every modification is done carefully and correctly.
- Plan on a conservative power increase – don't push it.

> *This book should be used in conjunction with a factory service manual, or a repair manual for your specific vehicle from GM, Chilton, or Haynes, which will include safety procedures, engine rebuilding information, torque specifications, cleaning, etc.*

9 Conclusions

You have reached the end of the "read through" part of this book. The appendixes include related information, some of which may be of interest to some readers. ☺ Here are two of particular importance.

- The Excel spreadsheet in Appendix A is very valuable in helping to select the right turbocharger. It is also valuable in making the calculations accurately... it is easy to make mistakes when doing these calculations by hand.

- Appendix D will be very valuable to readers whose engine has a carburetor.

This is very important. In order to make a reliable turbocharged engine at low cost, provide plenty of oil to the turbocharger, cool the turbocharger well, keep the boost level reasonably low, ensure the engine always runs rich under boost, and never allow ANY knocks caused by detonation.

In this book I have described many mistakes I have made. Without exception, these mistakes have cost me money and time. Sometimes I call them "lessons learned". I learn more from my mistakes than from my successes. I hope you can learn from MY mistakes, and avoid making the same mistakes yourself.

For some components (turbochargers, intercoolers, etc.) you can save money by using no-name parts and/or used parts, and for other components I recommend new or name-brand parts. This is an area where my experience may be beneficial to readers. More details on this subject can be found on page 31.

With these examples, keep in mind all cars are different. Each new system will have to be designed and tuned on its own. These examples are a starting place, not the exact final solution.

There is no limit to what a person can do, but it takes time, determination, patience, and common sense {sometimes "uncommon sense" when we make decisions to do things different from the vast majority of other people}. This is not the quick kit installation, and it does not result in the maximum horsepower. With a lot of research up front it can be done.

→ You can do it with a lot of work, but it won't be easy.

Most importantly: When you modify an engine, it will not be as reliable as a stock engine. Any modifications described in this book should be done on a car that is only used for racing, not a car that you depend on to get to work every day.

Don't do this to your daily driver.

Turbocharging Normally Aspirated Engines on a Budget

 An Excel Spreadsheet for Turbocharger Compressor Sizing

Turbocharger compressor sizing is just a matter of a little math. I have created an Excel spreadsheet that includes all of the formulas to make it easier to find the effect of making changes. This Excel spreadsheet can be downloaded from **WagonerEngineering.com**. Otherwise this spreadsheet is not very complicated, and it only takes a short time to type it in.

The spreadsheet calculates the volume and mass of air moving through the engine, air temperatures, as well as approximate torque and HP. The spreadsheet itself is shown below. Input parameters are entered in cells B2 through B12. All the calculations are performed in cells B15 through B26, and all formulas used in the spreadsheet are shown in column D.

	A	B	D
1		INPUTS	*Source: "High Performance Fieros" by Robert Greg Wagoner*
2	Outside Air Ambient Temp in deg F	60	
3	Altitude above Sea Level in ft	0	
4	Engine Volume in Liters	3.40	
5	Engine Compression Ratio	7.50	
6	Engine RPM for Maximum HP	5000	
7	Engine Volumetric Efficiency VE	0.700	*Typical range is 0.70 to 0.85.*
8	Engine Air to Fuel Ratio AFR	13.00	*Should be rich. Enter 13.0 to 13.5.*
9	Turbocharger Boost Pressure in psi (Gauge)	11.00	*Set this to 0 if there is no turbocharger.*
10	Turbocharger Compressor Efficiency CE	0.78	*Find CE on the compressor map.*
11	Intercooler Efficiency	0.71	*Set this to 0 if there is no intercooler.*
12	Intercooler Pressure Drop in psi	1.00	*Set this to 0 if there is no intercooler.*
13			
14	**X & Y VALUES FOR CE GRAPH**		formulas
15	X = Air Flow in LBS / minute	25.5	B21*B22
16	Y = Turbocharger Boost Pressure Ratio PR	1.75	(B9+B19)/B19
17			
18		**OUTPUTS**	
19	Outside Air Pressure in psia (Absolute)	14.7	14.7-B3*0.000494258
20	Boost Pressure Ratio after Intercooler	1.68	(B19+B9-B12)/B19
21	Air Flow in cubic ft / minute	333.7	0.0177516*B4*B6*B7*B10*B20/(B10+(1-B11)*(B16^0.283-1))
22	Air Density in LBS / cubic ft	0.076	2.7*B19/(B2+460)
23	Engine HP	231.2	298.2*B21*B22/B8*(1-1/B5^0.35)*(1-ABS(14.7-B8)/14.7)^2
24	Engine Torque in Ft-Lb	242.9	16500/pi()/B6*B23
25	Turbocharger Outlet Temperature in deg F	174.2	(B2+460)*(B16^0.283-1)/B10+B2
26	Intercooler Outlet Temperature in deg F	93.1	B25-B11*(B25-B2)

To use this Excel spreadsheet, it is necessary to enter values from the compressor graph for the turbocharger. The X and Y coordinates needed for the compressor map are calculated on lines 15 and 16. A compressor map for the Garrett T04E-"50" turbocharger is shown on the next page. The dashed lines show the X and Y coordinates from the Excel spreadsheet above, which meet where the compressor efficiency is very close to 78%. This is the value that gets entered into cell B10.

This spreadsheet makes it easy to see the effects of outside air temperature, compression ratio, an intercooler, and other factors on the engine's performance. As a reference point, with the values shown in the spreadsheet above, the spreadsheet predicts 231.2 HP. For example, if the altitude was 3000 feet instead of sea level, the Engine Power drops to 216.1 HP. As another example, if there was no intercooler, the Engine Power drops to 209.8 HP. For a system without an intercooler

Turbocharging Normally Aspirated Engines on a Budget

set the intercooler pressure drop to zero and also the intercooler efficiency to zero in the Excel spreadsheet.

This spreadsheet will work for normally aspirated engines as well. Without a turbocharger, just set the turbocharger boost to 0 in cell B10, and the turbocharger compressor efficiency to 1 in cell B11. Appendix C includes calculations of various normally aspirated engines using this spreadsheet.

The spreadsheet is based on formulas found in books in the references, combined with my own modifications and simplifications. The formulas in the Excel spreadsheet are based on principles and formulas described in Appendix B.

Compressor Maps and Compressor Efficiency

The compressor pumps air into the intake manifold. No pump is perfect because of friction, which is to say its efficiency must be less than 100%. Compressor Efficiency (CE) is the amount of power that goes into building pressure divided by the total power put into the compressor. An example compressor map is shown below. The sets of curves in the graph of interest are the efficiency curves. The surge limit and RPM limit give operating boundaries. Turbocharger operation must be kept within these boundaries or the turbocharger could be damaged. The manufacturer gives other RPM curves, but they are not important for this analysis.

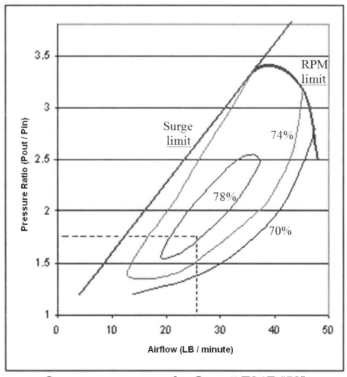

Compressor curves for Garrett T04E-"50"

The curves on these graphs show the Compressor Efficiency. The X-axis (bottom of the graph) shows the amount of air flowing through the compressor in LBS/minute at standard temperature and pressure (abbreviated STP). The term standard temperature and pressure denotes an exact reference temperature of 0 °C (273.15 K) and pressure of 100 kPa (1 bar). The left side of the graph shows the pressure ratio of compressor outlet pressure divided by its inlet pressure.

Turbocharging Normally Aspirated Engines on a Budget

The operating point from the example in the Excel spreadsheet above (X = 25.5, Y = 1.75) is plotted by dashed lines on the graph. By plotting the operating point on the graph, it is clear that the compressor efficiency is close to 78% at this operating point. This value is used as an input to the Excel spreadsheet into cell B10.

Pressure Ratio

Pressure Ratio is the outlet absolute pressure of the turbocharger's compressor compared to its inlet absolute pressure. This is calculated in cell B16 of the Excel spreadsheet. The outlet pressure is inlet pressure + boost pressure. The inlet pressure will be close to, but slightly below atmospheric pressure (14.7 PSI at sea level), due to restrictions in the plumbing and air cleaner at the turbocharger compressor inlet.

Altitude

The relationship between altitude and air pressure is not linear, but in a small range of altitude around sea level, a fairly accurate linear approximation can be made. The air pressure is 29.92 in/Hg at sea level. The air pressure has dropped to 24.89 in/Hg at 5000 feet. Using this ratio to adjust the air pressure at sea level from 14.7 PSIa, results in a derating factor of 0.000494258 PSIa per foot. This factor is used to get the air pressure based on altitude in cell B19 of the Excel spreadsheet.

Volumetric Efficiency

Volumetric Efficiency (VE) is a factor used to show the actual airflow divided by the ideal airflow based on cylinder volume. If the engine were a perfect air pump with no restrictions, it would move exactly the amount of air equal to the cylinder volume on each intake stroke. However it does not pump this much air because there are some restrictions in the intake ports and valves, and there will be some exhaust gas remaining in the cylinder at the end of the exhaust stroke, plus other factors such as camshaft timing. By adding the turbocharger, the intake and exhaust plumbing has been modified, and the flow is reduced. The volumetric efficiency is complicated because it actually changes with RPM. For example, the Volumetric Efficiency for the stock 2.8L Fiero engine is between a maximum of 0.831 at 3600 RPM down to 0.712 at 5200 RPM. See Appendix C for some examples. I have used VE = 0.7 in the example spreadsheet above.

Air-to-Fuel-Ratio (AFR)

For gasoline engines, an Air-to-Fuel ratio of 14.7:1 is almost ideal for burning all of the fuel and using up all of the oxygen in the air. Tuning a turbocharged engine will involve adjusting the AFR to be richer than 14.7:1, typically in the range of 13:1 at the relatively low levels of boost we are using in this book. By making the air/fuel mixture rich as the boost pressure increases, the excess fuel reduces the combustion chamber temperatures, which is beneficial for making more power for these turbocharger installations, but not good for fuel efficiency. Another drawback of making the air/fuel mixture too rich is that it tends to foul spark plugs and leave carbon deposits in the combustion chambers. Hence rich is good, but too rich is bad. If the air/fuel mixture goes lean, this increases combustion chamber and exhaust temperatures, which may result in detonation and damage the engine. Lean is always bad. For initial calculations with this spreadsheet enter 13 into cell B8 as the AFR. The spreadsheet can be updated later after the engine has been tuned and the actual AFR is known.

Turbocharging Normally Aspirated Engines on a Budget

Intercooler Parameters

An intercooler reduces the temperature of the air, after it has been compressed, before it reaches the engine. This temperature drop is calculated in cell B26 of the spreadsheet. Intercooler efficiency is used for calculations in cell B21 in the spreadsheet. Formulas for intercooler efficiency can be found in Appendix B. For the spreadsheet, if intercooler parameters have been provided or measured, use these values. Some examples are provided below. Intercooler efficiency is dependent on so many factors including its location in the car, and this makes the value difficult to know with certainty without actual measurements in each particular car. If the intercooler efficiency is not known, the best a person can do is guess. When I do not know the intercooler efficiency, I ordinarily start with an arbitrary number, such at 70% efficient, and then adjust it down to 60% efficient to see the impact on the performance.

This spreadsheet will work in a system without an intercooler too. If there is no intercooler, set the intercooler efficiency in cell B11 to 0 and the intercooler pressure drop in cell B12 to 0.

Explanation of Formulas used in the Excel spreadsheet

Compressor Volumetric Flow Equation

The basic equation below gives an approximation of the compressor airflow. In a four-stroke engine each cylinder fires every other revolution. The intake and exhaust open once every other revolution. So the engine moves an amount of air equal to its displacement once every 2 revolutions of the engine. This gets multiplied by volumetric efficiency, which is typically 80% to 85%, depending upon many factors about the engine such as intake restrictions, the particular camshaft, and flow. To be conservative use 70%; to be optimistic use 85%, and the real result will most likely fall somewhere between. The last factor is the total amount of pressure at the output of the compressor, scaled to atmospheric pressure, which is absolute pressure in BAR. For example, at STP (Standard Temperature and Pressure) the standard pressure at sea level is 14.7 PSI. This reference level is exactly 1 BAR. If, for example, the car is at sea level and the boost is exactly 14.7 PSI, then the total pressure is 29.4 PSI, which is exactly 2 BAR.

$$Compressor\_Airflow = 0.5 * RPM * D * VE * B / 1728$$

Where: Compressor_Airflow is calculated in cubic feet / minute (cfm)
RPM = maximum engine RPM
D = engine displacement in cubic inches (in^3)
VE = volumetric efficiency
B = absolute pressure in BAR

If the VE is changed from 0.70 to 0.85 in the spreadsheet, the calculated Engine Power increases significantly, from 301.2 HP to 365.7 HP. Considering the wide possible variation in VE, I have dropped other factors from the equations that make practically insignificant changes to the output, such as relative humidity.

Ideal Gas Law

The Ideal Gas Law relates air pressure, temperature, volume, and mass. The Ideal Gas Law is:

$$PV=nRT$$

Where: P is the absolute pressure (PSIa),
V is the air volume (cu ft)
n is air mass (LBS)
R is a constant = 0.37 $PSIa*ft^3/LBS/degR$
T is absolute temperature in degrees R (Rankine)
where degrees R = degrees F + 460.

This can be rearranged to solve for any of the variables if you know the other three. A conversion of outside air temperature from degrees F to Rankine is in line 22 of the Excel spreadsheet.

Compressor Efficiency

Compressor Efficiency (technically known as Adiabatic Efficiency) is the ratio of power input to a compressor that would have been required to compress a gas to a certain higher pressure divided by the actual power input to the compressor. To think of it another way, a compressor is less than 100% efficient, and the loss of heat during compression causes additional outlet air temperature rise from the turbocharger compressor. Most turbochargers will have compressor efficiency in the

Turbocharging Normally Aspirated Engines on a Budget

range of 65-80%. To figure the actual outlet temperature including the compressor efficiency, use the following formula.

IOTR / AE = AOTR
Where: *IOTR = Ideal Outlet Temperature Rise*
                 CE = Compressor (Compressor) Efficiency
                 AOTR = Actual Outlet Temperature Rise

This equation is included in the air temperature rise calculation below, which includes compressor efficiency.

Air Temperature Rise

The compressor temperature rise used in line 25 is calculated as follows.

Temperature_Rise = Temperature_in × [(Pressure_out/Pressure_in)^0.283-1]/CE
Where: *Temperature_Rise is temperature rise in degrees R*
                 Temperature_in is inlet temperature in degrees R
                 Pressure_out is absolute outlet pressure in PSIa
                 Pressure_in is absolute inlet pressure in PSIa
                 CE is Compressor Efficiency

Fuel Specific Heat Energy

Fuel Specific Heat Energy is important to these calculations because it is the amount of energy released in burning a particular fuel. This factor is buried within the equations in the Excel spreadsheets above.

Let's start by clearing up the terminology. I have found it misused frequently in Internet discussions.

- A *calorie* is the amount of heat required to raise the temperature of one gram of pure water by one degree Kelvin.

- *Heat Capacity* is the amount of heat required to raise the temperature of an object (or substance) by one degree Kelvin.

- *Specific Heat Capacity* or *Specific Heat* is the amount of heat required to raise the temperature of one gram of a substance by one degree Kelvin.

Almost all chemical reactions involve either the release of heat or the absorption of heat. The definitions above relate to heat absorption. Burning fuel involves the release of heat.

- *Fuel Specific Heat Energy* is the amount of energy released in burning a particular fuel.

The specific heat energy of the fuel is derived from the equation below.

Fuel Specific Heat Energy = LHV × MW / AFR
        Where: LHV is the lower heating value of the fuel (43 MJ/kg for gasoline)
                 MW is the molecular weight of air
                 AFR is the Air-to-Fuel ratio

Intercooler Parameters

An intercooler reduces the temperature of the air, after it has been compressed, before it reaches the engine. This temperature drop is calculated in line 26 of the spreadsheet. Intercooler efficiency is also used for calculations in line 21 in the spreadsheet. Intercooler efficiency can be calculated by the following formula.

IC efficiency % = [IC inlet temp - IC outlet temp] / [IC inlet temp – ambient temp]

Note these temperature measurements will be different based on many factors, particularly the external cooling airflow through the intercooler.

Fuel Vapor Pressure

The vapor pressure of a pure compound is the pressure exerted by its vapor in equilibrium with the liquid, as long as there is liquid remaining. However, fuels are not pure compounds, but are actually mixtures of many different hydrocarbons. Each of the different hydrocarbons in the fuel has a different vapor pressure. The vapor pressure of fuel is also difficult to measure. Finally fuel vapor pressure is also dependent upon temperature. These factors complicate any calculations that depend on fuel vapor pressure.

C VE Calculations for Various Engines

This appendix includes some sample results from the Excel spreadsheet presented in Appendix A. Recall I mentioned that the value for VE will be different for each type of engine, based on the particular airflow of that engine, and it is also dependent upon the exact RPM. VE is affected by even small changes in the airflow path. For example, adding a low restriction air filter will increase VE. On the other hand, a dirty air filter will reduce VE. As another example, if tuned headers were added, the VE would increase in the range of RPM where the headers are tuned. Tuned systems effectively have a resonance that increases the gain of the system over a range of RPM. It is possible for the VE to be greater than 1.0 in a range of RPM where the headers and intake are tuned to work the best.

That being said, typically the value of VE is within a range from 0.7 to 0.85 for older production automobiles. With the original small block Chevy engine, the only way to get VE to be 1.0 was to add a bunch of aftermarket parts, including replacing almost everything except the short block, such as heads, camshaft, intake, carb, and headers. This is also true for the 60° V6 used in Fieros. However, the newer production engines flow very well as built stock from the factory, as you will see in the subsequent calculations in this appendix.

Most production automobile engines have a relatively flat torque curve over the desired operating range. A flat torque curve comes from a flat VE curve. The two are directly related. For this reason, on a normally aspirated engine, the maximum VE is found at the same RPM as the maximum torque. Above and below this particular RPM the torque starts to fall. Similarly, above and below this RPM the VE also starts to fall. The maximum engine power is achieved above the RPM of the maximum torque, where the torque has fallen somewhat. Consequently the VE is always lower at the RPM for maximum HP than it is at the RPM where the maximum torque is found. The reason I mention this is because usually there is data for maximum HP at one particular RPM and maximum torque at a different RPM. This data can be used to get a good idea of the VE at these two RPMs.

In this Excel spreadsheet I have simplified the calculation to utilize a single value for VE. If a person has an idea of how the VE changes based on RPM, the VE parameter in the Excel spreadsheet can be adjusted based on RPM. These samples in this appendix show my estimation of VE for various engines.

In the calculations in this appendix I have kept the AFR = 13.5, to attempt to make equal comparisons between the engines. They may be running AFR a bit higher in these new engines to get more power for a given airflow, which would result in a slightly lower value of VE in these calculations. I have also kept the outside air temperature and altitude constant. If the temperature or altitude were higher, this would result in the calculated VE being higher. Changing any of these factors would lead to different results. I am sure that all of these VE numbers have some slight error, but the point of this is to make a comparison between the different engines.

Turbocharging Normally Aspirated Engines on a Budget

LS1

In the past I have been very interested in the 2000 Camaro with the LS1 engine, because that is the engine we put into a Fiero. The 2000 Camaro with the 5.7L LS1 engine is rated at 305 HP at 5200 RPM and 335 Ft-Lb at 4000 RPM, measured at the wheels. Adjusting this by a factor of 0.9 to account for gearbox loss to match the calculations on the next page, and working backwards, the VE calculated by this spreadsheet is 0.821 at 5200 RPM, as shown below.

Source: "High Performance Fieros" by Robert Greg Wagoner		
INPUTS		
Outside Air Ambient Temp in deg F	60	
Altitude above Sea Level in ft	0	
Engine Volume in Liters	5.67	
Engine Compression Ratio	10.10	
Engine RPM for Maximum HP	5200	
Engine Volumetric Efficiency VE	0.821	*Typical range is 0.70 to 0.85.*
Engine Air to Fuel Ratio AFR	13.50	*Should be rich. Enter 13.0 to 13.5.*
Turbocharger Boost Pressure in psi (Gauge)	0.00	*Set this to 0 if there is no turbocharger.*
Turbocharger Compressor Efficiency CE	1.00	*Find CE on the compressor map.*
Intercooler Efficiency	0.00	*Set this to 0 if there is no intercooler.*
Intercooler Pressure Drop in psi	0.00	*Set this to 0 if there is no intercooler.*
X & Y VALUES FOR CE GRAPH		*formulas*
X = Air Flow in LBS / minute	32.8	B21*B22
Y = Turbocharger Boost Pressure Ratio PR	1.00	(B9+B19)/B19
OUTPUTS		
Outside Air Pressure in psia (Absolute)	14.7	14.7-B3*0.000494258
Boost Pressure Ratio after Intercooler	1.00	(B19+B9-B12)/B19
Air Flow in cubic ft / minute	429.7	0.0177516*B4*B6*B7*B10*B20/(B10+(1-B11)*(B16^0.283-1))
Air Density in LBS / cubic ft	0.076	2.7*B19/(B2+460)
Engine HP	339.0	298.2*B21*B22/B8*(1-1/B5^0.35)*(1-ABS(14.7-B8)/14.7)^2
Engine Torque in Ft-Lb	342.4	16500/pi()*B6*B23
Turbocharger Outlet Temperature in deg F	60.0	(B2+460)*(B16^0.283-1)/B10+B2
Intercooler Outlet Temperature in deg F	60.0	B25-B11*(B25-B2)

Working backwards from the rated torque of 335 Ft-Lb at 4000 RPM, scaled up by the same factor we used before to account for gearbox loss, leads to a maximum VE of 0.892 at 4000 RPM. At its peak, the maximum VE of the 2000 Camaro LS1 is quite a bit lower than the peak of the LS6 engines. This is not surprising, because GM purposely used different camshafts, exhaust, etc. in the Camaro and in the Corvette, so that the Corvette would have higher torque and power than the Camaro.

The 2004 Corvette was offered with the 5.7L LS1 engine, rated at 350 HP at 5200 RPM and 375 Ft-Lb at 4000 RPM, measured at the wheels. If the calculations had been based on the 2004 LS1 Corvette instead of the 2000 LS1 Camaro, the VE results would be a lot closer to the LS6 engine. The 5.7L LS1 engine, rated at 350 HP at 5200 RPM, leads to VE of 0.941. The 5.7L LS1 engine, rated at 375 Ft-Lb at 4000 RPM, leads to VE of 0.998.

© 2012 Robert G. Wagoner
All rights reserved.

Turbocharging Normally Aspirated Engines on a Budget

Stock Fiero 2.8 L V6

Now let's look at the original 2.8L V6 engine that was stock in a Fiero. Back when Fieros were built, it was common to publish engine HP instead of wheel HP. The published ratings of a 2.8L V6 engine that came stock in a Fiero is 140 HP at 5200 RPM and 165 Ft-Lb at 3600 RPM, which roughly agrees with our own measurements of this engine, as described in my previous book. Since these are already measured at the engine, no scaling is necessary for gearbox efficiency. Working backwards the VE for this spreadsheet is 0.712 at 5200 RPM, as shown below.

	Source: "High Performance Fieros" by Robert Greg Wagoner	
	INPUTS	
Outside Air Ambient Temp in deg F	60	
Altitude above Sea Level in ft	0	
Engine Volume in Liters	2.80	
Engine Compression Ratio	8.90	
Engine RPM for Maximum HP	5200	
Engine Volumetric Efficiency VE	0.712	*Typical range is 0.70 to 0.85.*
Engine Air to Fuel Ratio AFR	13.50	*Should be rich. Enter 13.0 to 13.5.*
Turbocharger Boost Pressure in psi (Gauge)	0.00	*Set this to 0 if there is no turbocharger.*
Turbocharger Compressor Efficiency CE	1.00	*Find CE on the compressor map.*
Intercooler Efficiency	0.00	*Set this to 0 if there is no intercooler.*
Intercooler Pressure Drop in psi	0.00	*Set this to 0 if there is no intercooler.*
X & Y VALUES FOR CE GRAPH		*formulas*
X = Air Flow in LBS / minute	14.0	B21*B22
Y = Turbocharger Boost Pressure Ratio PR	1.00	(B9+B19)/B19
	OUTPUTS	
Outside Air Pressure in psia (Absolute)	14.7	14.7-B3*0.000494258
Boost Pressure Ratio after Intercooler	1.00	(B19+B9-B12)/B19
Air Flow in cubic ft / minute	184.0	0.0177516*B4*B6*B7*B10*B20/(B10+(1-B11)*(B16^0.283-1))
Air Density in LBS / cubic ft	0.076	2.7*B19/(B2+460)
Engine HP	139.9	298.2*B21*B22/B8*(1-1/B5^0.35)*(1-ABS(14.7-B8)/14.7)^2
Engine Torque in Ft-Lb	141.3	16500/pi()/B6*B23
Turbocharger Outlet Temperature in deg F	60.0	(B2+460)*(B16^0.283-1)/B10+B2
Intercooler Outlet Temperature in deg F	60.0	B25-B11*(B25-B2)

Working backwards from the rated torque of 165 Ft-Lb at 3600 RPM, leads to a VE of 0.831 at 3600 RPM. I would say this engine has asthma when compared to the other engines mentioned before. Not only is the maximum VE lower, it happens at a much lower RPM. It just doesn't breathe very well in comparison to the other engines in this appendix.

Turbocharging Carbureted Engines

It is possible to turbocharge a normally aspirated, carbureted engine, although it is not easy. Chapters 1 through 3 are applicable for information about sizing turbochargers and installing them. The Excel spreadsheet in Appendix A is just as applicable to carbureted engines as it is to fuel injected engines. The turbocharger system will need similar exhaust work, lubrication, cooling, wastegate control of boost pressure, and a Blow-Off Valve (BOV), as described in the previous chapters. An intercooler is still just as effective on a carbureted engine, and it can be installed immediately after the turbocharger, just as in fuel injected engines.

There is one advantage of using a carburetor instead of fuel injectors, which some people overlook. No tuning of the ECM, and no piggyback ECM, no MAF sensor calibration, etc. Simply install a larger carburetor. This can reduce the required effort significantly. Particularly for people with experience working on carbureted engines, this can be a powerful motivation to use a carburetor.

There are two methods for turbocharging a carbureted engine: Draw-Through and Blow-Through. The draw-through design puts the carburetor at the turbocharger inlet, to draw fresh air through the carburetor at atmospheric pressure, whereas in contrast, the blow-through design puts the turbocharger before the carburetor to pressurize the air before it goes to the carburetor.

The draw-through design is actually a bit easier to assemble in terms of not having to seal up the carburetor and not having to increase the fuel pressure. With a draw-through design, in general all that is necessary is to size the carburetor larger, based on the expected increase in airflow. However, there is a significant drawback of the draw-through design. The location of the carburetor is further from the intake manifold, and the plumbing through the turbocharger will undoubtedly be awkward, difficult, and it will cause flow restrictions and undesired turbulence in the intake manifold. In a draw-through system a BOV should be installed to avoid the vacuum that would occur when the turbocharger is spinning and the throttle is dropped shut. This BOV should be routed around the turbocharger in a closed loop, not vented to ambient air. I should mention the Excel spreadsheet in Appendix A will be close but not quite accurate in this particular application, because the turbocharger is pumping this denser air/fuel mixture instead of air only.

In the blow-through design, the turbocharger pressurizes the air before it goes through the carburetor. In order to make the carburetor work properly, it is necessary to seal the entire carburetor in a box, with sealed holes for the throttle shaft. Yes, I mean place the entire carburetor inside an airtight box. Ordinary fuel floats can crush under pressure, and special fuel floats filled with foam should be used, which are designed to work with the additional pressure. The fuel system needs to be modified with a 1:1 boost pressure regulator that can be referenced to the boost pressure, so the fuel pressure increases directly with the boost pressure. This will most likely require replacing the original mechanical fuel pumps with a high volume electric fuel pump.

Don't pay attention to internet lore that suggests extreme jet changes to the carburetor. Instead pick a carburetor sized for the new airflow.

With the extra information in this section, a custom turbocharger system can be designed for a carbureted engine based on the other information in this book.

Turbocharging a 5.7L LS1 V-8 Engine in a C5 Corvette

This is a planned future project, and therefore there are no photographs at this time, only a dream.

The LS1 / LS6 engine is GM's next generation of V8, known as Gen III. The LS1 supersedes and improves upon the original small block V8 and the LT1 series in every way. The LS1 has an aluminum block and aluminum cylinder heads, significantly reducing the weight as compared to the older V8 engines with a steel block. The camshaft is hollow to reduce weight. Even the intake manifold is lightweight, being made out of a composite plastic. This engine has many of the features previously considered to be aftermarket power adders in the earlier generation V8 engines. It has hydraulic roller lifters that allow the camshaft lobes to have fast ramp rates. The block was designed to be stiffer and have lower vibration than all the older V8 engines. The oil pan is made of cast aluminum, and it contributes to the rigidity of this engine. The main caps have six bolts, four vertical and two horizontal. The connecting rods are the strongest ones ever used in a production V8 engine, manufactured with tight tolerances in weight at both ends. The stock LS1 heads flow better than many ported small block Chevy heads, and have higher combustion chamber efficiency. This list was only a small portion of the improvements in the LS1 / LS6.

One or Two Turbochargers

The closer the turbocharger is to the engine exhaust, the better, in terms of providing the most available energy to the turbocharger turbine wheels. Physically, on the LS1 (or any other V-8 engine), since there are 4 cylinders on one side and 4 cylinders on the other side, the ideal solution is to use two turbochargers, one on each side, positioned close to the exhaust manifold. In contrast, if a single turbocharger is used, it should be located at a point where all of the exhaust comes together into one pipe, which is quite a ways from the engine, and very inconvenient with dual exhaust.

For drivability it is better for the turbochargers to spin up and provide the desired boost at lower RPM. In terms of turbo lag, two smaller turbochargers will have less rotating mass, and will spin up faster than one large turbocharger. To offset this, the size a single turbocharger can be reduced so it will spin up faster, but that has a negative impact on overall horsepower. Either way you look at it, the twin turbocharger system will outperform a single turbocharger system.

In terms of cost, a single turbocharger will be the lower cost solution. It seems obvious without much explanation, and this is confirmed by multiple sources on-line and in the references. Recall, from the beginning this book has been dedicated to low cost turbocharger installations with minimal intrusion and highest reliability, not highest horsepower. So if cost is the most important issue, a single turbocharger should be used. However, this is a Corvette, and twin turbochargers is my preference, even though it will not result in the lowest cost. My initial cost estimates indicate this setup will exceed the $1000 goal for the other cars in this book... the total cost may be in the neighborhood of $1500.

LS1 or LS6

Primarily, the choice between LS1 or LS6 is also cost driven. The cost of the Corvette with an LS6 engine is significantly higher than the Corvette with an LS6 engine, even though the performance is not that much greater. Another important factor is the compression ratio. In general engines with lower compression ratio can take more boost pressure before detonation. The compression ratio of the LS1 is 10.25:1, and the compression ratio of the LS6 is 10.5:1, so I would expect the LS1 to allow higher boost pressure before detonation. Hence the PLAN is to start with an LS1.

Turbocharging Normally Aspirated Engines on a Budget

Camshaft Selection

There are numerous articles in magazines and on the internet with examples of turbocharging an LS1 engine. As described in the camshaft section of the Appendix, a more radical camshaft profile will improve the power over the stock camshaft profiles, while simultaneously reducing low end torque slightly. For example, based on a number of articles with actual measured horsepower and torque values, a Comp 281LRHR13 camshaft will increase the horsepower by about 5% in either a normally aspirated LS1 engine as well as a mildly turbocharged LS1 engine. While this is certainly nothing to sneeze at, remember the goal of this book is to get the most performance for the lowest cost, and this kind of a camshaft change does not make the cut. Besides, in terms of reliability, I find the more I mess with, the less reliable it gets. Consequently I plan to keep the stock camshaft.

Turbocharger System Block Diagram

The planned system configuration is shown below. Notice both turbocharges feed together into a single front mounted intercooler, then to the BOV, which is placed before the MAF sensor. This allows the BOV to vent directly to the outside air through its own air filter. Swoosh!

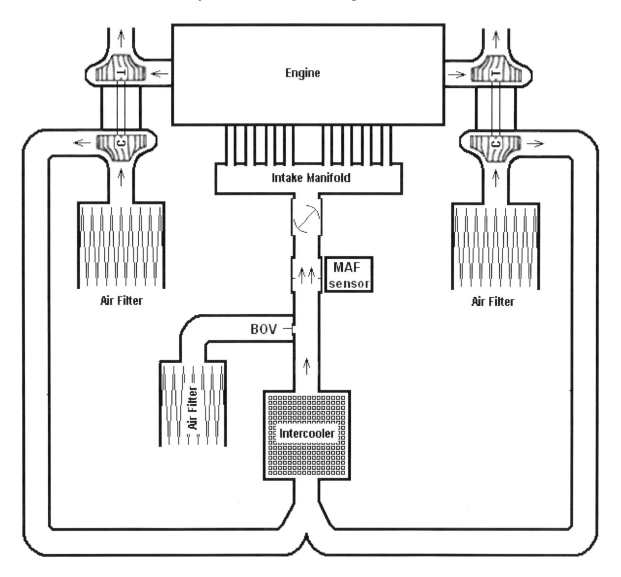

Turbocharging Normally Aspirated Engines on a Budget

Reprogramming the ECM

Tuning will be accomplished with "LS1 Edit". The creators of LS1 Edit deserve a lot of credit for making this program "user friendly". We used this software to change many parameters of the LS1 PCM when we installed an LS1 into a Fiero, and I plan to use the same software again for this project.

Excel Spreadsheet Calculations of Airflow and Horsepower

The calculation of expected power is 467 HP with 7 PSI boost, as shown in the spreadsheet below.

	INPUTS	Source: "High Performance Fieros" by Robert Greg Wagoner
Outside Air Ambient Temp in deg F	60	
Altitude above Sea Level in ft	0	
Engine Volume in Liters	5.67	
Engine Compression Ratio	10.25	
Engine RPM for Maximum HP	5200	
Engine Volumetric Efficiency VE	0.844	Typical range is 0.70 to 0.85.
Engine Air to Fuel Ratio AFR	13.00	Should be rich. Enter 13.0 to 13.5.
Turbocharger Boost Pressure in psi (Gauge)	7.00	Set this to 0 if there is no turbocharger.
Turbocharger Compressor Efficiency CE	0.70	Find CE on the compressor map.
Intercooler Efficiency	0.75	Set this to 0 if there is no intercooler.
Intercooler Pressure Drop in psi	0.50	Set this to 0 if there is no intercooler.
X & Y VALUES FOR CE GRAPH		formulas
X = Air Flow in LBS / minute	46.7	B21*B22
Y = Turbocharger Boost Pressure Ratio PR	1.48	(B9+B19)/B19
OUTPUTS		
Outside Air Pressure in psia (Absolute)	14.7	14.7-B3*0.000494258
Boost Pressure Ratio after Intercooler	1.44	(B19+B9-B12)/B19
Air Flow in cubic ft / minute	611.6	0.0177516*B4*B6*B7*B10*B20/(B10+(1-B11)*(B16^0.283-1))
Air Density in LBS / cubic ft	0.076	2.7*B19/(B2+460)
Engine HP	466.6	298.2*B21*B22/B8*(1-1/B5^0.35)*(1-ABS(14.7-B8)/14.7)^2
Engine Torque in Ft-Lb	471.3	16500/pi()/B6*B23
Turbocharger Outlet Temperature in deg F	146.6	(B2+460)*(B16^0.283-1)/B10+B2
Intercooler Outlet Temperature in deg F	81.6	B25-B11*(B25-B2)

Turbocharger Selection

Using two turbochargers to make 467 HP requires each one to handle 234 HP. Running in parallel, the total airflow through the engine will be split between the two turbochargers. These can be small turbochargers, similar to the ones used in the MR2 Spyder (page 58), for example.

In a future revision of this book, I hope to include photographs of this setup. ☺

F Rebuilding Turbochargers

Like some other things I have described in this book, I am providing this information so you can learn from my mistakes. I suggest you DO NOT rebuild a worn-out turbocharger. We have found this the hard way; it is less expensive and more reliable to purchase a new no-name brand turbocharger. The biggest problems we have discovered with used turbochargers are damaged impellers, trying to get it balanced after it is reassembled, and worn shafts. The paragraphs below describe how Nick and I rebuilt a turbocharger with a worn shaft. I am not going to describe the procedure for rebuilding, but just the worn shaft problem and our solution.

We purchased a used Garrett T3 turbocharger that needed to be rebuilt. Our T3 turbocharger had too much wear where the bearings hold the shaft. The diameter of the shaft in the area of the bearing wear was 0.398" OD, and the diameter of the shaft was 0.399" OD where the bearing had not rubbed the shaft. These dimensions and locations are indicated in the photograph below.

We purchased a turbocharger rebuild kit made specifically for this turbocharger. The standard bearing measures 0.401" ID. When reassembled with the standard bearing, the end of the shaft could move by about 0.010" which is too much end play.

My son, Nick, used a lathe to turn down and polish the shaft to 0.398" OD across the entire bearing surface. We also purchased an extra set of undersized bearings, and turned them down to an inner diameter of 0.399" ID. These new bearing have an inside surface that is wider than the original bearings. When reassembled the end of the shaft could only move by about 0.002".

My recommendation: Do not attempt to rebuild a turbocharger yourself. If you plan to use a rebuilt turbocharger, get one rebuilt by a professional that comes with a guarantee.

G. Camshafts for Turbocharged Engines

As described in chapter 2, to minimize cost in low boost applications, I recommend keeping the stock camshaft. However, just to be complete, I want to discuss camshafts and dispel some misinformation found on the internet.

First a disclaimer: This section refers to camshafts with fixed valve timing. Variable valve timing allows different camshaft profiles to be used at different RPMs. Variable valve timing provides such flexibility that the engineers at the engine manufacturer could tune a camshaft profile that is highly optimized for a particular turbocharger setup. Optimizing camshaft profiles for engines with variable valve timing are beyond the comprehension of the author.

Camshafts have valve overlap, a time when both intake and exhaust valves are open simultaneously. All engines work best with some overlap. Since turbochargers cause an exhaust restriction and raise the exhaust back pressure, it is possible for the exhaust pressure to be higher than intake pressure during valve overlap, when both valves are open at the same time. If this happens, the airflow between the valves would reverse, driving exhaust gas back into the intake. Because of this, valve overlap needs to be less with a turbocharged engine than with a normally aspirated engine. A normally aspirated engine can tolerate wider valve overlap and wider durations of both intake and exhaust valves, and using a camshaft designed for a normally aspirated car will result in less than optimum performance on a turbocharged engine.

> **Calculating Valve Overlap**
>
> Valve Overlap = Exhaust Closing + Intake Opening
>
> Example Comp Cams XR276 HR
> - Intake Opening (IO) = 32 deg BTDC
> - Intake Closing (IC) = 64 deg ABDC
> - Exhaust Opening (EO) = 75 deg BBDC
> - Exhaust Closing (EC) = 27 deg ATDC
>
> Valve Overlap = 27 deg + 32 deg = 59 deg

Contrary to older literature, mild cam profiles are not necessarily better for a turbocharged engine than for a normally aspirated engine. In general anything that adds power to a normally aspirated motor will also add power on a turbocharged engine. Camshaft lobe separation angle (LSA) is a factor that has been considered important to turbocharged engines in the past. Generic recommendations can be found in older literature for camshaft of LSA = 114 degrees, but these generic recommendations are misleading at best and sometimes just plain incorrect. Valve overlap and the other camshaft factors are more important. Camshaft profiles that minimize valve overlap will not out-power traditional cam profiles with more valve overlap. A camshaft with LSA = 114 degrees might be fine for a full race engine that has a "radical" camshaft, designed to provide high horsepower at 6000 RPM, but such an engine will not be optimized for low end torque. Lower values of LSA, such as LSA = 110 degrees, actually work better in many cases. Hence the LSA will not be discussed further herein.

Keep in mind that smaller turbochargers tend to create more back pressure at lower RPM, allowing them to create boost at lower RPM. This is an important consideration to camshaft selection. A larger turbocharger, which has less exhaust back pressure, works better with a camshaft designed for a normally aspirated engine than a smaller turbocharger. In order to optimize a system with a smaller turbocharger, changing the camshaft is more important. It is also important to note that older turbochargers were less efficient than modern turbochargers, and the exhaust back pressure is lower with newer high-efficiency turbochargers. Older turbochargers require the exhaust valve to close earlier than newer high-efficiency turbochargers. Similar to the discussion about small vs.

large turbochargers, a newer high-efficiency turbochargers, which has less exhaust back pressure, works better with a camshaft designed for a normally aspirated engine than an older turbocharger.

Another important factor in camshaft selection is the cylinder head. A good flowing head allows a camshaft with shorter duration to work better at higher RPM, for both normally aspirated and turbocharged engines. And just like a normally aspirated engine, a turbocharged engine will have more peak HP at high RPM with a "radical" cam, and it will have more torque at lower RPM with a "mild" cam. Effectively all the factors related to flow and RPM range that go into designing a normally aspirated engine still apply to a turbocharged engine. If the goal is to build the highest performance engine possible, free-flowing heads should be used, and the camshaft can be selected to take advantage of the free-flowing heads. However, examples in this book will always use the stock heads in order to keep the total cost of the project lower.

All this information about camshafts being said so far has not yet resulted in any specific recommendations on camshaft selection. Remember this book is all about turbocharging normally aspirated engines on a budget. The goal is to get the most "bang for the buck", which requires carefully selecting the components that are changed. With that in mind, none of the examples in this book will touch the bottom end, including the camshaft. The turbochargers are sized accordingly, and the boost pressures are limited to relatively low levels, so that the camshaft does not have to be replaced. This will result in an engine with good torque at lower RPM, and still a significant power increase too. Recall from the introduction this book will not focus on ultimate HP, nor will it cover high dollar race engines, so we are not trying to come up with a "radical" race cam. I make these assumptions because that is the way I prefer my engines to run. I like an engine with a lot of torque at the bottom end, so I lean towards a mild camshaft. With moderate levels of boost the stock camshaft for the normally aspirated engine works just fine for a turbocharged engine.

If you do decide to change the camshaft, I recommend selecting an "off-the-shelf" camshaft designed specifically for a turbocharged engine. Each camshaft manufacturer will have their own recommendations for a particular engine, and certainly specific recommendations from the manufacturer are better than any generic recommendation for camshaft durations and LSA such as those found in generic tables on the internet. This is particularly true when considering ramp rates can be quite different between two different camshafts, and the ramp rate can have a very significant impact on the performance given particular opening and closing times. When changing a camshaft, also change the other related components, such as lifters, springs, etc. as recommended by the camshaft manufacturer.

Superchargers vs. Turbochargers

When it comes to forced induction systems, there is always a debate about: "Which is better—a supercharger or a turbocharger?". Both have advantages and disadvantages. This section is an attempt to make a fair comparison between supercharging and turbocharging. Many of the concepts in this book will apply equally to both supercharging and turbocharging. However, the specific examples of actual installations in this book are turbocharged only.

There is an abundance of new and used turbochargers available in the market, much more than superchargers, so the average price of the turbocharger will be lower. The total system cost of installing either a supercharger or a turbocharger onto a normally aspirated engine may be similar in some cases if everything is purchased new including the intake manifold. However, in cases where a turbocharger is bolted on, using the existing intake manifold and throttle body, the total system cost is less with a turbocharger than with a supercharger, particularly when comparing to a roots-style supercharger.

Besides cost, performance is another important factor. Turbocharger lag is one of its greatest drawbacks. A turbocharger is driven by exhaust gas, and when starting from a slow speed, due to its inertial mass, a turbocharger needs time to speed up (spool up) before the compressor can create much boost. This is a bigger problem for large turbochargers, because smaller turbochargers spool up quicker, and this reduces the lag. Besides lag, another drawback of a turbocharger is they are not able to generate much boost at low RPM because they are not positive displacement pumps. The centrifugal supercharger is similar in this respect. In contrast, positive displacement superchargers create boost over a much wider range of RPM. These are the primary reasons some people prefer superchargers over turbochargers on racecars.

A turbocharger is more efficient than a supercharger. There are a lot of factors that go into this, and the difference is not as easy to identify as some would like to make it. A supercharger draws power directly from a pulley on the crankshaft, which takes power directly from the engine, power that could have gone directly to the transmission. The erroneous typical argument is that a turbocharger is driven by energy in the exhaust gasses that would otherwise be lost in the exhaust. This is only partially true. This is a more complicated analysis than it seems, because it takes energy to turn the turbine to turn the compressor, and this energy comes from the exhaust gas, that exhaust gas being driven directly from the piston, resisting the piston as it moves up on the exhaust stroke. This power, which is used to spin the turbine of a turbocharger, effectively comes from the crankshaft, except through the piston instead of directly from a pulley on the crankshaft as it would for a supercharger. Hence both a turbocharger and a supercharger take power from the crankshaft to spin the turbine. The factors that make the difference include pump efficiencies, turbine efficiencies, gearbox efficiency, camshaft timing, exhaust and intake flow, temperatures, intercoolers, etc. Some people find it surprising that the wastegate operation is a significant factor in the efficiency of a turbocharger: when the wastegate is open, the turbocharger does not create the extra exhaust back pressure, so it does not take extra power from the engine. The final proof to the efficiency debate becomes clear when looking at a completely different piece of evidence: cars and trucks designed for maximum fuel efficiency utilize turbochargers and not superchargers.

Superchargers do not disrupt the exhaust gas flow, and a supercharged system can benefit from the additional power available from a tuned and free-flowing exhaust. A turbocharger does create additional exhaust back pressure, and this negatively impacts overall output power of the engine. Keep in mind a turbocharger's wastegate partially bypasses this exhaust flow restriction during normal driving conditions where the boost is not needed. A turbocharger has the potential of producing more peak horsepower than superchargers, because of the higher efficiency with a turbocharger, and because a turbocharger can make massive boost pressures at high RPM.

There are other contrasts between a supercharged system and a turbocharged system. Certain brands of centrifugal superchargers are known to have a loud gear noise (whine), whereas a turbocharger is ordinarily quieter than the supercharger. Also a turbocharger tends to quiet the exhaust noise. A supercharger has a wider usable RPM band, with high torque at lower RPM than a turbocharger. There are many other factors, and clearly both have advantages and disadvantages. This book will focus on how to turbocharge an engine without attempting to answer the question, "Which is better - a supercharger or a turbocharger?", because the author believes there is no clear answer to that question.

References organized by the author's preference

	Title	Author	Publisher
1	High Performance Fieros, 3.4L V6, Turbocharging, LS1 V8, Nitrous Oxide	Robert Wagoner	Lulu
2	Turbo: Real-World High-Performance Turbocharger Systems	Jay K. Miller	SA Design
3	Maximum Boost: Designing, Testing, and Installing Turbocharger Systems	Corky Bell	Motorbooks
4	Turbochargers HP49	Hugh MacInnes	H.P. Books
5	How to Tune and Modify Engine Management Systems	Jeff Hartman	Motorbooks
	Websites		
6	http://www.airspacemag.com/history-of-flight/climb.html		
7	http://en.turbolader.net/Technology/History.aspx		

Turbocharging Normally Aspirated Engines on a Budget

Index

AFR (Air-to-Fuel Ratio), 13, 18, 24, 31, 41, 42, 48, 50, 51, 56, 60, 67, 68, 79, 102, 105, 107
Air Conditioner, 34
air filter, 11, 13, 14, 22, 30, 32, 34, 41, 48, 53, 75, 76, 82, 86, 94, 95, 97, 107
Air Flow, 5, 20, 48, 66, 86, 101
Air Mass, 104
Air Volume, 104
Altitude, 7, 9, 66, 100, 102, 107
Bearing, bearings, 17, 18, 31, 89, 114
BLM (Block Learn Multiplier), 41, 42
boost creep, 23, 58, 74, 77
BPC (Base Pulse Constant), 42
BSFC (Brake Specific Fuel Consumption), 4, 12, 22, 39
Calorie, 105
camshaft, 4-6, 25, 34, 71, 86, 94, 102, 104, 107, 108, 116, 117
carburetor, 6, 9, 99, 110
CE (Compressor Efficiency), 12, 100-102, 104, 105
clutch, 4, 16, 24-26, 40, 94
Compression, compression ratio, 4, 18, 37, 38, 71, 79, 86, 100, 104, 111
Coolant, Cooling, Water, 5, 14, 17, 19, 20, 34, 49, 58, 76, 95, 105, 106, 110
detonation, 4, 9, 13, 14, 18, 19, 24, 38, 56, 102, 111
drivetrain, 4, 9, 16, 25, 26, 79, 83
EPROM (Erasable Programmable Read Only Memory), 41, 42
fan, 12, 19, 20, 40
fuel pump, 4, 21, 30, 38, 39, 49-51, 72, 94, 110
Garrett, 28, 29, 31, 32, 45, 91, 93, 100, 101, 114
head spacer, 18, 37, 38
heat capacity, 13, 15, 105
Ideal Gas Law, 104
ignition, 5, 13, 14, 18, 19, 24, 28, 30, 37, 47, 50, 51, 60, 72, 75

knock, 4, 9, 13, 14, 18, 24, 28, 37, 41, 50
LS1, 5, 6, 9, 108, 111-113
LS6, 108, 111
Mandrel Bent Tubing, 4, 33
MAF (Mass Air Flow) Sensor, 5, 14, 21-23, 48-51, 53, 55, 56, 59, 61, 63, 66, 67, 72, 79, 82, 86
MAP (Manifold Absolute Pressure) sensor, 4, 14, 30, 35, 36, 42-45, 66, 67, 69, 72, 94, 96
Nitrous Oxide, 6, 22
Octane, 13, 18
OX Sensor, Oxygen Sensor, 24, 30, 32, 50, 52, 53, 62, 65, 102
piggyback FIC, 5, 14, 24, 47, 49-52, 59, 67, 72
piston, 14, 18, 28, 35, 45, 117
rebuild, 3, 5, 28, 31, 45, 91, 114
reliability, reliable, 3, -5, 7, 9, 17-25, 28, 31, 39, 45, 50, 51, 67, 77, 79, 83, 91, 111, 112
rev limiter, 42, 75
spark knock, see knock
spark plug, 18, 19, 37, 102
specific heat, 13, 15, 105
spring, springs, 23, 38, 57, 67, 68, 74, 95, 116
STP (Standard Temperature & Pressure), 101, 104
Subaru, 6, 20, 67
supercharger, supercharging, 4, 5, 7, 11-13, 15, 117, 118
temperature, 12-15, 17-19, 21, 22, 37, 41, 66, 100-107, 117
transaxle, 32, 40
transmission, 22, 25, 74, 89, 94, 117
turbine, 11, 13, 15, 17, 23, 58, 111, 117
valve covers, 34
vapor pressure, 106
VE (Volumetric Efficiency), 35, 36, 42, 96, 102, 104
wastegate actuator, 23, 67, 68
weight, 4, 7, 9, 10, 105, 111
wire, wiring, 17, 33, 37, 51, 52

Made in the USA
San Bernardino, CA
01 May 2016